Molecular Plant Pathology, Volume I

Three week loan

Please

The Practical Approach Series

SERIES EDITORS

D. RICKWOOD
Department of Biology, University of Essex
Wivenhoe Park, Colchester, Essex CO4 3SQ, UK

B. D. HAMES
Department of Biochemistry and Molecular Biology, University of Leeds
Leeds LS2 9JT, UK

Affinity Chromatography
Anaerobic Microbiology
Animal Cell Culture
Animal Virus Pathogenesis
Antibodies I and II
Biochemical Toxicology
Biological Membranes
Biosensors
Carbohydrate Analysis
Cell Growth and Division
Cellular Calcium
Cellular Neurobiology
Centrifugation (2nd Edition)
Clinical Immunology
Computers in Microbiology
Crystallization of Nucleic Acids and Proteins
Cytokines
The Cytoskeleton
Directed Mutagenesis
DNA Cloning I, II, and III
Drosophila
Electron Microscopy in Biology
Electron Microscopy in Molecular Biology
Enzyme Assays
Essential Molecular Biology I and II
Fermentation
Flow Cytometry
Gel Electrophoresis of Nucleic Acids (2nd Edition)
Gel Electrophoresis of Proteins (2nd Edition)
Genome Analysis
HPLC of Small Molecules
HPLC of Macromolecules
Human Cytogenetics
Human Genetic Diseases
Immobilised Cells and Enzymes
Iodinated Density Gradient Media
Light Microscopy in Biology
Liposomes
Lymphocytes
Lymphokines and Interferons
Mammalian Development

Molecular Plant Pathology

Volume I
A Practical Approach

Edited by

S. J. GURR

and

M. J. McPHERSON

Centre for Plant Biochemistry and Biotechnology
Department of Biochemistry and Molecular Biology
University of Leeds

and

D. J. BOWLES

Centre for Plant Biochemistry and Biotechnology
Department of Biochemistry and Molecular Biology
and Department of Pure and Applied Biology
University of Leeds

OXFORD UNIVERSITY PRESS
Oxford New York Tokyo

Oxford University Press
Walton Street, Oxford OX2 6DP

Oxford is a trade mark of Oxford University Press

Published in the United States
by Oxford University Press, New York

A catalogue record for this book
is available from the British Library

Library of Congress Cataloging in Publication Data
Molecular plant pathology: a practical approach/edited by Sarah
Jane Gurr, Michael J. McPherson, and Dianna J. Bowles.
(Practical approach series)
Includes bibliographical references and index.
1. Plant diseases—Molecular aspects—Handbooks, manuals, etc.
2. Phytopathogenic microorganisms—Molecular aspects—Handbooks,
manuals, etc.   3. Plant-pathogen relationships—Molecular aspects—
Handbooks, manuals, etc.   4. Molecular microbiology—Technique—
Handbooks, manuals, etc.   5. Plant molecular biology—Technique—
Handbooks, manuals, etc.   I. Gurr, Sarah Jane.   II. McPherson, M.
J.   III. Bowles, Dianna J.   IV. Series.
SB732.65.M65   1992 581.2—dc20   91-24886
[ISBN 0–19–963103–4]hbk.
[ISBN 0–19–963102–6]pbk.

Typeset by Cotswold Typesetting Ltd, Gloucester
Printed in Great Britain by
Information Press Ltd, Eynsham, Oxon

Preface

Molecular plant pathology is one of the fastest moving and most exciting fields in biology and has directly benefited from advances in modern molecular techniques. These have been applied to both pathogen and plant, allowing us to develop some understanding of the organisms themselves and of the complex interactions leading to compatibility or incompatibility.

The intention has been to provide a comprehensive handbook describing both the latest molecular techniques and more 'classical' approaches. Coverage is also given to areas of research that we believe will become increasingly important in molecular plant pathology: the nature of the signalling molecules involved in the recognition between plant and pathogen, the nature of the signal transduction pathways in the plant that lead from those initial recognition events to defence-related changes in gene expression, and strategies for the isolation of disease resistance genes.

The contents encompass the expertise of a broad range of research workers with hands-on experience in plant pathology. Some contributors specialize in specific classes of pathogens, whilst others focus on the diverse responses of the plant to pathogen invasion.

Due to the sheer volume of information required to provide this handbook, it has proved necessary to produce two *Practical Approach* volumes. The contents follow a sequence incorporating techniques for specific pathogens, followed by plant responses at the levels of genes, proteins, and defence-related compounds, the elicitor molecules and signalling pathways, and finally the disease resistance genes. The contents of the books are organized into six sections, the first three of which contain a number of chapters describing complementary methods. A broad range of protocols are provided from pathogen isolation and culture, through physiology and biochemistry to molecular biology and techniques for the localization of genes and their products *in situ* within cells of the infected plant.

Within Volume I, Section 1 provides an introduction to four classes of pathogen: viruses, bacteria, fungi, and nematodes. The importance of the first three to plant pathologists is obvious. However it is now increasingly recognized that nematodes are crucially important pathogens of a wide range of crop species, yet our molecular understanding of their interaction with their host plants is only just beginning. We hope the inclusion of nematodes in this handbook will lead to the wider recognition of these parasites in molecular plant pathology.

Whilst detailed methods for various molecular biology approaches are provided in Section 2, the reader is also referred elsewhere for further discussion of techniques in plant molecular biology (1), polymerase chain reaction applications (2, 3), and a range of other general techniques in gel electrophoresis

(4, 5) and nucleic acid methodology (6, 7). Similarly, protocols for carbohydrate analyses are complementary to those found in (8). A chapter on the use of baculoviruses as expression vectors has been included to highlight their enormous potential for the study of gene expression in host–pathogen interactions.

Within Volume II, Section 3 provides detailed protocols for the analysis of gene products induced during defence responses. Sections 4 and 5 concentrate, respectively, on the preparation of elicitors and on the analysis of signal transduction pathways. Descriptions of methods for the determination of calcium levels, phosphorylation, and membrane inositide turnover should both provide a readily accessible set of techniques for use by plant pathologists and also be of use to other plant biologists who may not have extensive biochemical experience.

The two volume set ends with Section 6, which provides a review of strategies for the isolation of disease-resistance genes, a challenge that represents the dominant research objective of many molecular plant pathologists. Whilst no resistance gene has yet been isolated, techniques such as those described in Chapter 27 should hold the key to success in this search: success that will represent a landmark in plant science.

We hope these volumes will be of use to the community of plant pathologists world-wide and will provide a ready source of important methods and approaches that can be applied to the study of plant and pathogen, whatever the specialized interest of the individual investigator.

SARAH JANE GURR

Leeds   MICHAEL J. McPHERSON

August 1991   DIANNA J. BOWLES

References

1. Shaw, C. S. (ed.) (1989). *Plant molecular biology: a practical approach*. IRL, Oxford.
2. McPherson, M., Quirke, P., and Taylor, G. R. (ed.) (1991). *PCR: a practical approach*. IRL, Oxford.
3. Erlich, M. A. (ed.) (1989). *PCR technology, principles and applications for DNA amplification*. Stockton Press, New York.
4. Hames, B. D. and Rickwood, D. (ed.) (1990). *Gel electrophoresis of proteins: a practical approach* (2nd edn). IRL, Oxford.
5. Hames, B. D. and Rickwood, D. (ed.) (1990). *Gel electrophoresis of nucleic acids: a practical approach* (2nd edn). IRL, Oxford.
6. Sambrook, J., Fritsch, E. F., and Maniatis, T. (ed.) (1989). *Molecular cloning: a laboratory manual* (2nd edn). Cold Spring Harbour Press, New York.
7. Davis, L. G., Dibner, M. D., and Batten, J. F. (ed.) (1986). *Basic methods in molecular biology*. Elsevier, New York.
8. Chaplin, M. F. and Kennedy, J. F. (ed.) (1987). *Carbohydrate analysis: a practical approach*. IRL, Oxford.

Contents

11. The polymerase chain reaction 123

Michael J. McPherson, Richard J. Oliver, and
Sarah Jane Gurr

12. Analysis of defence gene transcriptional regulation 147

Maria J. Harrison, Arvind D. Choudhary, Michael A. Lawton, Christopher J. Lamb, and Richard A. Dixon

13. In situ hybridization in plants 163

David Jackson

Appendices

Contents

Contributors to Volumes I and II

PETER ALBERSHEIM
Complex Carbohydrate Research Centre, University of Georgia, PO Box 5677, Athens, Georgia 30613, USA.

PAUL AHLQUIST
Institute for Molecular Biology, K25 Linder Drive, University of Wisconsin, Madison, Wisconsin 53706, USA.

HOWARD J. ATKINSON
Centre for Plant Biochemistry and Biotechnology, Department of Pure and Applied Biology, University of Leeds, Leeds, LS2 9JT, UK.

CARL BERGMANN
Complex Carbohydrate Research Centre, University of Georgia, PO Box 5677, Athens, Georgia 30613, USA.

DAVID H. L. BISHOP
NERC Institute of Virology and Environmental Microbiology, Mansfield Road, University of Oxford, Oxford, OX1 3SR, UK.

THOMAS BOLLER
Plant Physiology Department, University of Basel, Habel Street, Basel 4056, Switzerland.

DIANNA J. BOWLES
Centre for Plant Biochemistry and Biotechnology, Department of Biochemistry and Molecular Biology and Department of Pure and Applied Biology, University of Leeds, Leeds, LS2 9JT, UK.

DOUGLAS S. BUSH
Department of Botany, 2017 Life Sciences Building, University of California, Berkeley, CA 94720, USA.

IAIN R. CAMERON
NERC Institute of Virology and Environmental Microbiology, Mansfield Road, University of Oxford, Oxford, OX1 3SR, UK.

JONG-JOO CHEONG
Complex Carbohydrate Research Centre, University of Georgia, PO Box 5677, Athens, Georgia 30613, USA.

ARVIND D. CHOUDHARY
Botany Department, Nagpur University Campus, Nagpur 440010, India.

JOHN M. CLARKSON
School of Biological Sciences, University of Bath, Claverton Down, Bath, BA2 7AY, UK.

DONALD A. COOKSEY
Department of Plant Pathology, University of California, Riverside, California, USA.

MICHAEL J. DANIELS
Sainsbury Laboratory, John Innes Centre, Colney Lane, Norwich, NR4 7UH, UK.

ALAN DARVILL
Complex Carbohydrate Research Centre, University of Georgia, PO Box 5677, Athens, Georgia 30613, USA.

ALAN P. DAWSON
School of Biological Sciences, University of East Anglia, Norwich, NR4 7TJ, UK.

RICHARD A. DIXON
Plant Biology Division, Samuel Roberts Noble Foundation Inc., PO Box 2180, Ardmore, Oklahoma 73402, USA.

BJORN DRØBAK
Department of Cell Biology, John Innes Institute, Colney Lane, Norwich, NR4 7UH, UK.

ROBERT EDWARDS
Plant Biology Division, Samuel Roberts Noble Foundation Inc., PO Box 2180, Ardmore, Oklahoma 73402, USA.

IAN B. FERGUSON
Mount Albert Research Centre D.S.1R Private Bag Auckland NZ.

ROBERT DE FEYTER
Department of Plant Pathology, University of Florida, Gainesville, Florida 32611, USA.

MARC G. FORTIN
Department of Plant Science, McGill University, St-Anne de Bellevue, Quebec, Canada H9X 1CO.

DAVID M. FRANCIS
Department of Vegetable Crops, University of California, Davis CA 95616, USA.

JOHN FRIEND
Department of Applied Biology, University of Hull, Hull, HU6 7RX, UK.

STEPHEN FRY
Botany Department, King's Buildings, University of Edinburgh, Edinburgh, EH9 3JH, Scotland, U.K.

DEAN W. GABRIEL
Department of Plant Pathology, University of Florida, Gainesville, Florida 32611, USA.

SARAH JANE GURR
Centre for Plant Biochemistry and Biotechnology, Department of Biochemistry and Molecular Biology, University of Leeds, Leeds, LS2 9JT, UK.

MICHAEL HAHN
Complex Carbohydrate Research Centre, University of Georgia, PO Box 5677, Athens, Georgia 30613, USA.

KIM E. HAMMOND-KOSACK
Sainsbury Laboratory, John Innes Centre, Colney Lane, Norwich, NR4 7UH, UK.

JOHN HARGREAVES
Department of Agricultural Sciences, University of Bristol (AFRC), Long Ashton Research Station, Long Ashton, Bristol, BS18 9AP, UK.

MARIA J. HARRISON
Plant Biology Division, Samuel Roberts Noble Foundation Inc., PO Box 2180, Ardmore, Oklahoma 73402, USA.

ROGER HULL
John Innes Virus Research Centre, Colney Lane, Norwich, NR4 7UH, UK.

DAVID JACKSON
Department of Cell Biology, John Innes Institute, Colney Lane, Norwich, NR4 7UH, UK.

RUSSELL L. JONES
Botany Department, 2017 Life Sciences Building, University of California, Berkeley CA 94720, USA.

Contributors

HEINRICH KAUSS
University Kaiserslautern, Fachbereich Biologie, 1075 Kaiserslaustern, Germany.

NOEL T. KEEN
Department of Plant Pathology, University of California, Riverside, California, USA.

RICHARD V. KESSELI
Department of Vegetable Crops, University of California, Davis, CA 95616, USA.

HELMUT KESSMAN
CIBA-GEIGY, Department of Agriculture, CH4002, Basle, Switzerland.

ALAN KOLLER
Complex Carbohydrate Research Centre, University of Georgia, PO Box 5677, Athens, Georgia 30613, USA.

PHILIP KRONER
The Blood Research Institute, The Blood Centre of South Wisconsin, 1701 West Wisconsin Avenue, Milwaukee, Wisconsin 53233, USA.

CHRISTOPHER J. LAMB
Plant Biology Laboratory, Salk Institute for Biological Studies, 10010 North Torry Pines Road, La Jolla, California 92138, USA.

MICHAEL A. LAWTON
AgBiotech, Rutgers, The State University of New Jersey, PO Box 231, New Brunswick, New Jersey, USA.

RICHARD LEEGOOD
Department of Animal and Plant Sciences, University of Sheffield, Sheffield, S10 2TN, UK.

VENG-MENG LO
Complex Carbohydrate Research Centre, University of Georgia, PO Box 5677, Athens, Georgia 30613, USA.

MICHAEL J. McPHERSON
Centre for Plant Biochemistry and Biotechnology, Department of Biochemistry and Molecular Biology, University of Leeds, Leeds, LS2 9JT, UK.

Contributors

RICHARD MICHELMORE
Department of Vegetable Crops, University of California, Davis, California 95616, USA.

PHILIP MULLINEAUX
Department of Applied Genetics, John Innes Institute, Colney Lane, Norwich, NR4 7UH, UK.

RICHARD J. OLIVER
Norwich Molecular Plant Pathology Group, School of Biological Science, University of East Anglia, Norwich, NR4 7TJ, UK.

ILAN PARAN
Department of Vegetable Crops, University of California, Davis, California 95616, USA.

KEITH ROBERTS
Department of Cell Biology, John Innes Institute, Colney Lane, Norwich, NR4 7UH, UK.

JULIE D. SCHOLES
Department of Animal and Plant Sciences, University of Sheffield, Sheffield, S10 2TN, UK.

HAO SHEN
Department of Plant Pathology, University of California, Riverside, California, USA.

DAVID THRELFALL
Department of Applied Biology, University of Hull, Hull, HU6 7RX, UK.

GEOFFREY TURNER
Department of Molecular Biology and Biotechnology, University of Sheffield, Sheffield, S10 2TN, UK.

KATE VAN DEN BOSCH
Department of Biology, Texas A and M University, College Station, Texas 77843, USA.

IAN WHITEHEAD
Department of Applied Biology, University of Hull, Hull HU6 7RX, UK.

CHANG-HSIEN YANG
Department of Vegetable Crops, University of California, Davis, California 95616, USA.

Abbreviations

ARS	autonomously replicating sequence
BMV	brome mosaic virus
BSA	bovine serum albumin
βME	2-mercaptoethanol
CAT	chloramphenicol acetyl transferase
CCMV	cowpea chlorotic mottle virus
CsCl	caesium chloride
cDNA	complementary DNA
CTAB	cetyl trimethyl ammonium bromide
DEPC	diethylpyrocarbonate
DIG	Digoxigenin
DMSO	dimethylsulphoxide
dNTP	deoxyribonucleotide triphosphate
ds DNA	double-stranded DNA
ds RNA	double-stranded RNA
DTT	dithiothreitol
ECL	enhanced chemiluminescence
EDTA	ethylenediamine tetraacetic acid
EGTA	ethyleneglycol aminoethyl ether tetraacetic acid
ELISA	enzyme-linked immunosorbent assay
EMS	ethyl methane sulphonate
EtBr	ethidium bromide
EtOH	ethanol
FCS	fetal calf serum
GLC	gas–liquid chromatography
GRA	gel retardation assay
GuHCl	guanidium hydrochloride
HEPES	hydroxyethylpiperazine ethanol sulphonic acid
HeBS	Hepes buffered saline
HPLC	high pressure liquid chromatography
HPRNI	human placental ribonuclease inhibitor
IPTG	isopropyl-β-D-thiogalactoside
LMP	low melting point
MeOH	methanol
MES	N-morpholine ethanol sulphonic acid
MLOs	mycoplasma-like organisms
MLV	murine leukaemia virus
mRNA	messenger RNA
NaAc	sodium acetate

NPV	nuclear polyhedrosis virus
n.t.	nucleotide
ORF	open reading frame
PAGE	polyacrylamide gel electrophoresis
PBS	phosphate buffered saline
PCN	potato cyst nematode
PCR	polymerase chain reaction
PEG	polyethylene glycol
pfu	plaque forming unit
PMSF	phenylmethylsulphonyl fluoride
pV	pathovar
PVP	polyvinylpyrrolidone
RAPD	random amplified polymorphic DNA
RFLP	restriction fragment length polymorphism
RKN	root-knot nematode
rRNA	ribosomal RNA
RNAse A	ribonuclease A
rpm	revolutions per minute
RT	reverse transcriptase
SCN	soybean cyst nematode
SDS	sodium dodecyl sulphate
SDW	sterile distilled water
ssDNA	single-stranded DNA
SS phenol	salt-saturated phenol
TAE	Tris-acetate-EDTA
Taq	*Thermus aquaticus* polymerase
TBE	Tris-borate EDTA
TE	Tris-EDTA
TLC	thin layer chromatography
TMV	tobacco mosaic virus
tRNA	transfer RNA
UPGMA	unweighted pair group-method with averaging
UV	ultraviolet
WDV	wheat dwarf virus

1

Viruses

ROGER HULL

1. Introduction

Viruses are ideal subjects for study using molecular biological techniques because of their relative simplicity. This section describes both the range of known plant viruses and how to purify and characterize a new virus.

1.1 What is a virus?

Viruses are obligate parasites; the infectious units comprise either RNA or DNA enclosed in a protective coat of virus-coded protein. Some viruses also have lipids in their coat.

Viruses cause a range of plant diseases (for a review see ref. 1). The most common symptoms of these diseases are mosaics or mottles of dark green, light green, and sometimes yellow. The pattern of mosaics or mottles is usually regulated by the distribution of veins on the leaf; thus in monocotyledonous species the symptoms are frequently longitudinal stripes following the parallel venation.

Plant viruses differ in the numbers of plant species that they can systemically infect. Some have wide host ranges, e.g. alfalfa mosaic virus can infect at least 430 species in 51 families; others have narrow host ranges, e.g. rice necrosis virus is reported to infect only rice. However, the recognition that viruses can cause subliminal infections in apparent non-hosts (i.e. replicate in the initially infected cells but not spread to surrounding cells) raises the question of exactly what is a host. In this section we consider a host to be a plant species in which a virus gives a full systemic infection.

Plant viruses are transmitted naturally in a variety of specific means. For infection a virus has to enter through damage of the cuticle and cellulose cell walls; there is no evidence for cell-surface receptors for plant viruses. In nature this damage can be inflicted either mechanically, e.g. through breakage of leaf hairs, or more commonly, biologically by an insect, fungus, nematode, or pollen vector. The interactions between the virus and its vector are often very specific.

In the laboratory many plant viruses are introduced into their hosts mechanically. This is usually effected by grinding up an infected leaf in an appropriate buffer, adding an abrasive such as celite (diatomaceous earth) or carborundum and gently rubbing the extract on the surface of the leaf of the host plant. Systemic hosts usually give the same symptoms on mechanical inoculation as those induced on vector inoculation. Other plant species may develop necrotic local lesions (hypersensitive response) in which the cells around the initial point of entry die, frequently limiting the virus to a few cells. There are other viruses which cannot be transmitted mechanically. For these one has to use the natural vector or, in some cases, grafting or transfer by dodder (*Cuscuta* spp.). For details of how to transmit plant viruses either mechanically or by insect vector or grafting see ref. 2.

1.2 Groups of plant viruses

Knowledge of the shape of the virus particles and the nature of the viral genome, together with other information (e.g. vector, cytopathology) may enable the virus to be assigned to a particular group. It is then often possible to make various extrapolations, about, for example, genome organization and expression mechanisms.

There are about 650 known plant viruses, most of which are classified into 34 groups (see *Figure 1*). Details of the properties of members of these groups can be found in refs 3–10. About 75% of plant viruses, including most economically important pathogens, have genomes of plus-strand RNA. Others have genomes of minus-strand RNA, double-stranded RNA, single-stranded DNA, or double-stranded DNA encapsidated in their particles. One unusual feature of some groups of plant viruses is that the viral genome is divided between two segments (e.g. tobraviruses, comoviruses) or three segments (e.g. hordeiviruses, cucumoviruses) which are encapsidated separately; such viruses are termed *multicomponent*. These virus components can usually be separated by rate zonal and/or isopycnic centrifugation.

The variety of shapes of the particles of plant viruses covers the range from isometric (spherical), rigid and flexuous rods, bacilliform to bullet-shaped. It is important to determine the shape and stability of the particles early in the study of a new virus as these properties will have implications for the choice of purification techniques. Electron microscopy of crude samples from infected plants (see ref. 2) may elucidate both the shape of the particle and give an indication of particle stability by the effect of the negative stain on the particles. For instance, many salt labile viruses are unstable in sodium phosphotungstate at neutral pH but are stable in uranyl acetate. However, viruses which occur in low concentrations, and especially those with isometric particles can be difficult to visualize and to differentiate from ribosomes by electron microscopy. In these cases one has to try and assign the virus to a group on other properties and extrapolate the information on the particles. A good example of this situation is the luteoviruses, which are aphid-transmitted in the circulative manner, are not mechanically transmitted, frequently cause yellowing symptoms, and show serological relationships between several viruses (11). From these properties it

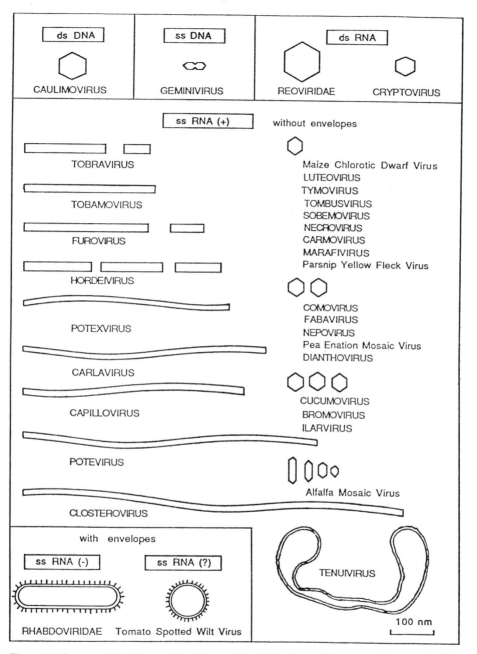

Figure 1. Groups of plant viruses. The diagram shows the shape of the particle and the genome type of the 34 recognized groups. Courtesy of R. I. B. Francki.

3

can be suggested that the particles would be isometric and that they would be restricted to the phloem.

2. Virus purification

2.1 Propagation of plant viruses

Plant hosts are used in the study of molecular aspects of a given virus for four reasons:

- growing the virus up for purification;
- separating the virus from contaminating viruses or strains;
- assessing the infectivity of virus preparations;
- retaining the virus culture.

The host used for virus purification should be:

- susceptible to systemic infection;
- one in which the virus reaches a high titre;
- horticulturally soft (i.e. produces much sap);
- should not contain undesirable substances (i.e. large amounts of polyphenol oxidases or fibres);
- easy to grow and be free from possible contamination with other viruses.

It is desirable early in the study of a new virus to use techniques such as enzyme-linked immunosorbent assay (ELISA) (12) or local-lesion assay to compare the titres of the virus in potential purification hosts and to determine the optimum time for harvesting. During the selection of the best host it is advisable to ascertain the optimum growing conditions for it to produce the virus. For instance, it is best to grow certain isolates of barley yellow mosaic virus in cereals at relatively low temperatures (e.g. below 20°C). Usually the optimum conditions for virus production are those under which the host is making most rapid growth.

Viruses can be separated from contaminating viruses or strains either by using different hosts or by the use of local-lesion hosts, since it is thought that each local lesion represents a single infection. Lesion counts can also be used to assess the concentration of infectious virus particles. However, for sap samples the possibility of inhibitors must be considered (2, 13). For the maintenance of virus cultures one should use a host that is

- long-lived;
- not severely affected by the virus;
- unlikely to be contaminated by other viruses;
- easy to propagate vegetatively.

With most viruses which are easily transmitted mechanically it is usually better to store the inoculum as dried leaf material as this will prevent any 'genetic drift' of

4

the virus. Infected leaf material is dried effectively by cutting the leaves into narrow strips and storing them over dehydrated calcium chloride. The dried material can be ground up in the appropriate buffer and used as inoculum. It is possible to preserve many viruses for 10 years or more in this way.

2.2 Purification

There is no universal technique for purifying plant viruses and for each virus there is at least one preferred technique, if not more. Detailed purification procedures for many viruses are described in refs 14 and 15. However, there are certain basic features to be taken into account in designing a virus purification protocol and some of these are described in ref. 16. It is useful to have a procedure for monitoring virus concentrations at various stages of purification and to assess the effects of various agents on the virus. ELISA is invaluable for monitoring, if an antiserum is available, otherwise techniques such as local-lesion assay or electron microscopy can be used. Whichever procedure is developed there are essentially four steps to obtaining a purified virus preparation:

1. Extraction from the plant (see Section 2.2.1).
2. Clarification of the extract (see Section 2.2.2).
3. Concentration of the virus (see Section 2.2.3).
4. Final purification (see Section 2.2.4).

2.2.1 Extraction from the plant

The first step is to break up the cell walls and release the cell contents. Damage to the virus particles by enzymes released during this step can be controlled, at least to a certain extent, by the use of buffers to preserve the pH, the addition of substances to inhibit enzymes, and the use of low temperatures (0–5°C).

There are various ways of breaking up plant tissues. The most frequently used instrument is the blender which can be either top- or bottom-driven. The efficiency of a blender is determined by the speed at which the blades rotate, the shape of the blades, and the shape of the vessel. However, it may be difficult to blend fibrous material satisfactorily, especially with a top-driven blender and certain blenders can cause the shearing of long flexuous virus particles (17). In these cases, and in cases where the virus is restricted to the vascular tissue, it is often more effective to freeze the plant tissue in liquid nitrogen and grind it to a fine powder before the addition of the buffer. The digestion of cell walls with cellulases is often effective at releasing phloem-limited viruses such as luteoviruses (11). A buffer is chosen in which the virus particles are known to be stable and in which they do not precipitate. The isometric particles of some viruses, e.g. bromoviruses, swell at alkaline pHs making the viral RNA susceptible to nucleases, whereas those of other viruses, e.g. cucumoviruses, may precipitate at pHs between 5 and 6. Elongated rod-shaped particles (e.g. those of poteviruses) tend to aggregate at neutral pHs and buffers with pHs around 8 to 9 and of moderate ionic strength should be used. Acidic buffers, with a pH slightly below

5, are useful as normal plant proteins are precipitated at low pH. Various additives to the extraction buffer are used primarily to inhibit enzyme action but they may also break up macromolecules other than virus particles. The most common problem is polyphenol oxidases which 'tan' the virus coat protein. These can be inhibited by reducing reagents (e.g. 0.5% 2-mercaptoethanol, 20 mM sodium sulphite), but in severe cases of polyphenol oxidase activity materials such as hide powder can be used to 'capture' the enzyme. Chelating compounds can be used to reduce enzyme activity and to destabilize ribosomes and phytoferritin. However, the particles of some viruses, e.g. sobemoviruses, are stabilized by divalent cations; in these cases chelators should be avoided. The particles of some viruses, e.g. caulimoviruses, are contained in proteinaceous inclusion bodies from which they must be released by, for example, overnight treatment with 2.5% Triton X-100 plus 1.5 M urea.

2.2.2 Clarifying the extracts

The virus particles must be separated from the normal constituents of cells such as membranes, ribosomes, proteins, and nucleic acids. If the virus particles stay in solution, organelles and large membrane fragments can usually be removed by low speed centrifugation (up to 10 000 g for 10 min). Emulsification with organic solvents, such as butanol or chloroform, was previously used to remove many contaminants, however, safety considerations have resulted in this approach being less frequently used. In most cases good preparations of virus can be obtained without the use of organic solvents. The addition of Triton X-100 often prevents particles aggregating to membranous materials and may lead to cleaner preparations.

2.2.3 Concentrating the virus

The process of concentrating the virus from large volumes of dilute solutions from the initial extracts provides further ways in which to remove contaminants. There are two common ways in which virus particles are concentrated:

(a) High-speed centrifugation (e.g. at 60 000–150 000 g for 2–3 h depending upon the sedimentation coefficient of the virus). The particles can often be separated from many normal host proteins and nucleic acids. Ribosomes and phytoferritin, which have similar sedimentation coefficients to those of viruses, can be disrupted by chelating agents or by moderately-high salt concentrations, provided, of course, that these treatments do not affect the virus particles. Centrifugation through a cushion of 5–10% sucrose often leads to a cleaner virus preparation.

(b) Precipitation using polyethylene glycol (PEG). The precipitability of a virus by PEG is dependent upon, amongst other factors, the shape of the virus particle and the concentrations of PEG and of salt. As a guide, rod-shaped particles can usually be precipitated by 2.5% w/v PEG (mol. wt 6000), 0.1 M NaCl; and isometric particles by 10% w/v PEG, 0.1 M NaCl.

2.2.4 Final purification

Viruses prepared to the concentration stage are not usually pure enough for many molecular techniques. However, it must be remembered that cloning is a purification technique and, if this is the aim, partially purified preparations of virus can be used as sources of nucleic acid.

The further purification of viruses is usually by rate-zonal and/or isopycnic centrifugation (16, 18). Among the factors to be taken into account in choosing a procedure are the following:

(a) Rate-zonal centrifugation results in the dilution of the virus band whereas isopycnic centrifugation leads to the concentration of the band.

(b) Isopycnic centrifugation in chaotropic salts, such as caesium chloride, can remove contaminating proteins and nucleic acids adhering to virus particles. However, several viruses are unstable in caesium chloride but may be stable in caesium sulphate.

(c) Rate-zonal centrifugation is usually performed in swinging, bucket rotors. Isopycnic centrifugation is best done in angle rotors which generate the gradient more quickly and give better band separation on gradient reorientation when the rotor stops.

(d) The materials for rate-zonal gradients are usually less expensive than those for isopycnic gradients.

3. Characterization of viruses

The basic information that one needs for molecular biological studies on a plant virus is the composition of the virus particles which, as noted in Section 1.2, comprise either RNA or DNA encapsidated in coat protein. The coat of most viruses is made up of subunits of a single species of virus-encoded protein. It is also useful to have information on other viral gene products, and this section outlines some of the procedures used.

3.1 Viral nucleic acids

Viral nucleic acids can be studied following extraction from purified virus particles. The basic methods involve either disrupting the bonds which maintain the particle structure, denaturing the coat protein, or enzymatically digesting the coat protein (16). Once extracted, the form of nucleic acid can be determined by enzymatic treatments coupled with gel electrophoresis (16) and the use of restriction endonucleases to determine if the DNA is double-stranded (19). Gel electrophoresis (16, 20) will often define the number of nucleic acid components which are encapsidated. However, two points must be considered when interpreting gel electrophoresis patterns.

(a) Subgenomic RNAs may be encapsidated and thus the number of genomic

7

RNA species could be overestimated. The best way to determine the composition of a divided genome is to isolate the nucleic acid species separated for example, by gel electrophoresis, and to test the infectivity of various combinations of them.

(b) The number of nucleic acid species can be underestimated if two of them have the same mobility in gels. This can be recognized by anomalies between viruses which have been assigned to the same group on other criteria. For unassigned viruses it may become apparent when trying to relate cDNA clones to RNA species.

3.2 Viral coat proteins

The number and size of coat-protein species is usually determined by gel electrophoresis of denatured proteins (16, 21). This procedure assumes that both the viral and molecular-weight marker proteins are totally denatured and have the same net charge per unit molecular weight. Some viral coat proteins, e.g. those of potexviruses and caulimoviruses, show atypical migration in gels; this can be detected by the use of Ferguson plots (22). Another problem can be the degradation of the coat protein by proteases which can result in two or more bands. The use of different purification methods to produce different proportions of the protein bands should alert one to this problem.

3.3 Non-structural proteins

These proteins are often very difficult to detect. However, with some viruses, e.g. poteviruses, non-structural proteins accumulate in infected cells to form characteristic aggregation bodies which can be detected by electron microscopy of thin sections. Some of these aggregation bodies have been purified and antisera raised to them (23). Recently, immunogold labelling of thin sections, viewed by electron microscopy, has proved to be useful in detecting non-structural proteins. The antisera can be raised to oligopeptides predicted from the nucleic acid sequence of the viral genome. For some methods relevant to these procedures see refs 24 and 25.

References

1. Matthews, R. E. F. (1991). *Plant virology* (3rd edn). Academic Press, New York.
2. Hill, S. A. (1984). *Methods in plant virology*, Vol. 1. Blackwell, Oxford.
3. Matthews, R. E. F. (1982). *Intervirology*, **17**, 173.
4. Francki, R. I. B., Milne, R. G., and Hatta, T. (1985). *Atlas of plant viruses*, Vol. 1. CRC Press, Boca Raton, Florida.
5. Francki, R. I. B., Milne, R. G., and Hatta, T. (1985). *Atlas of plant viruses*, Vol. 2. CRC Press, Boca Raton, Florida.
6. Francki, R. I. B. (ed.) (1985). *The plant viruses*, Vol. 1, *Polyhedral viruses with tripartite genomes*. Plenum, New York.

7. van Regenmortel, M. H. V. (ed.) (1986). *The plant viruses*, Vol. 2, *The rod-shaped plant viruses*. Plenum, New York.

8. Koenig, R. (ed.) (1988). *The plant viruses*, Vol. 3, *Polyhedral viruses with monopartite RNA genomes*. Plenum, New York.

9. Milne, R. C. (ed.) (1988). *The plant viruses*, Vol. 4, *The filamentous plant viruses*. Plenum, New York.

10. Hull, R., Brown, F., and Payne, C. (1989). *Virology: directory and dictionary of animal, bacterial and plant viruses*. Macmillan, London.

11. Casper, R. (1988). In *The plant viruses* Vol. 3, *Polyhedral viruses with bipartite genomes* (ed. R. Koenig), p. 235. Plenum, New York.

12. Clark, M. F. and Bar-Joseph, M. (1984). In *Methods in virology* Vol. 7 (ed. K. Maramorosch and H. Koprowski), p. 51. Academic, New York.

13. Bawden, F. C. (1964). *Plant viruses and virus diseases* (4th edn). Ronald, New York.

14. Murant, A. F. and Harrison, B. D. (ed.) (1970 to date). CMI/AAB and recently AAB descriptions of plant viruses Nos 1–354.

15. Kurstak, E. (ed.) (1982). *Handbook of plant virus infections*. Elsevier/North Holland Biomedical Press, Amsterdam.

16. Hull, R. (1985). In *Virology: a practical approach* (ed. B. W. J. Mahy), p. 1. IRL, Oxford.

17. Bar-Joseph, M. and Hull, R. (1974). *Virology*, **62**, 522.

18. Rickwood, D. (ed.) (1984). *Centrifugation: a practical approach* (2nd edn). IRL, Oxford.

19. Maniatis, R., Fritsch, E. F., and Sambrook, J. (1982). *Molecular cloning: a laboratory manual*. Cold Spring Harbor Laboratory Press, New York.

20. Rickwood, D. and Hames, B. D. (ed.) (1982). *Gel electrophoresis of nucleic acids: a practical approach*. IRL, Oxford.

21. Hames, B. D. and Rickwood, D. (ed.) (1981). *Gel electrophoresis of proteins: a practical approach*. IRL, Oxford.

22. Hedrick, J. L. and Smith, A. J. (1968). *Arch. Biochem. Biophys.* **126**, 155.

23. Dougherty, W. G. and Hiebert, E. (1980). *Virology*, **104**, 174.

24. Hills, G. J., Plaskitt, K. A., Young, N. D., Dunigan, D. D., Watts, J. W., Wilson, T. M. A. *et al.* (1987). *Virology*, **161**, 488.

25. Martelli, G. P. and Russo, M. (1984). In *Methods in virology*, Vol. 7 (ed. K. Maramorosch and K. Koprowski), p. 143. Academic Press, New York.

2

Geminiviruses (wheat dwarf virus)

PHILIP MULLINEAUX

1. Introduction

Episomal (i.e. autonomously replicating) vectors derived from animal DNA viral sequences, and which can be maintained either transiently or permanently in cultured cells, are widely used in studies of transcriptional regulation in animals. The geminiviruses are a group of plant DNA viruses (for reviews see refs 1 and 2) from which episomal vectors for plant cells could be developed. Three properties of these viruses are relevant for vector development.

(a) The predominant unencapsidated form of virus DNA which can be extracted from infected tissue, accumulates in nuclei as a supercoiled double-stranded (ds) molecule of c. 2.7 kb.

(b) The coat protein gene can be deleted without affecting dsDNA replication. In the monopartite geminiviruses, such deletion mutants are not capable of symptom formation on, or systemic spread through, whole plants (3). These deleted virus genomes can be termed 'disarmed'.

(c) In our experience with wheat dwarf virus (WDV), no limit to the size of inserted foreign DNA into the WDV genome has been observed; this contrasts with the caulimoviruses. Although WDV genomes containing inserts can replicate as ds forms, they are not capable of initiating infections in whole plants.

This chapter focuses on exploitation of the ability of the WDV genome to replicate autonomously in inoculated cells and protoplasts of the Gramineae. It also describes some of the routine procedures used in our laboratory for studying this phenomenon. In addition, the reader should be aware that, in theory, the principles outlined in this section could be applied to any plant species within the host range of a geminivirus. In this respect, the wide host range of this virus group is highly relevant (1). Potentially, the most universal application of geminivirus replicons will be to improve the ease, reproducibility, and sensitivity of transient expression assays for the quantification and analysis of transcriptional regulatory sequences. This may involve:

- elevation of the copy number of an introduced sequence;
- maintenance of its expression over several days (or perhaps weeks);

- the possibility of employing intact cell inoculation techniques, other than protoplasts.

The ability of a geminivirus-based vector to replicate to a high copy number may indeed compensate for the relative inefficiency of intact-cell inoculation.

Routine preparation of viable protoplasts from the tissues of many species, especially from members of the Gramineae, can be technically very difficult. As a consequence, many transient expression experiments are conducted using protoplasts prepared from tobacco mesophyll cells or suspension cultures, which may not be the ideal choice. In addition, many laboratories which do not possess expertise in the preparation and use of protoplasts find it difficult to establish these procedures. For these reasons there are considerable advantages in developing 'low-tech' approaches to the transient assay of promoters and other regulatory sequences in plants. Development of geminivirus episomal vectors, coupled to procedures for the inoculation of different tissues, is therefore highly desirable. Our first attempts to develop a non-protoplast inoculation procedure will be described in this chapter.

2. Vectors based on the WDV replicon

Several convenient cloning vehicles for the insertion of foreign DNA into the WDV genome are being developed in this laboratory. A typical example is that of pWDV2 (*Figure 1*). Several features of this plasmid are worth noting.

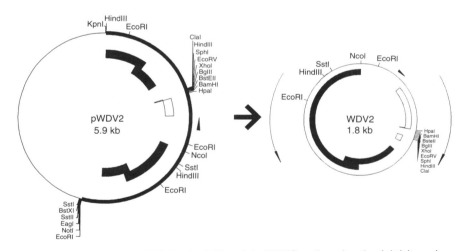

Figure 1. The plasmid pWDV2 (on the left) and the WDV2 replicon (on the right) it produces upon inoculation into *Triticum monococcum* protoplasts. The black arcs are the complementary-sense open reading frames (ORFs) C1 and C2, which are required for WDV dsDNA replication (4). The white arc is the virion sense ORF, V1. The arrowhead (▶) on both plasmids indicates the position of the virion sense promoter. The arrows on the WDV2 circle indicate the direction of transcription of the ORFs.

- The coat protein gene (V2 open reading frame; ORF) has been deleted, rendering this virus DNA incapable of initiating a plant infection.

- The C1 and C2 ORFs are retained and are required for dsDNA replication (4) along with a putative origin of replication, possibly located in the region between the 5′ ends of the V1 and C1 ORFs.

- There are four unique restriction sites located at the deletion junction of the V2 ORF to facilitate the insertion of foreign sequences.

- There is a direct repeat of *c.* 50% of WDV DNA either side of the deleted V2 ORF. The repeated sequences facilitate escape, via recombination of the virus DNA plus insert from the plasmid construct, upon entry into the nucleus of an inoculated cell or protoplast.

pWDV2 is based on a Bluescript™ vector (5), is ampicillin resistant, and should be maintained in *rec*A *Escherichia coli* (*E. coli*) strains (e.g. HB101; ref. 6). This, and similar plasmids containing tandem repeats of virus DNA, should be checked periodically to ensure continued integrity of the construct.

3. Replication of WDV DNA in *Triticum monococcum* protoplasts

3.1 The suspension culture and preparation of protoplasts

The non-regenerable cell suspension culture of *Triticum monococcum* was first described for use in gene transfer experiments by Lorz *et al.* (7). The rapid growth rates of suspension cultures and relatively high percentage of protoplasts which undergo sustained cell division (compared with many suspension cultures of the Gramineae) have allowed us to develop procedures for observing the replication and expression of the WDV genome. It should be noted that any WDV-based construct which can replicate in *T. monococcum* protoplasts can be used subsequently in other whole cell inoculation procedures. Methods for maintenance of *T. monococcum* suspension cultures and the preparation of protoplasts derived from these cell suspensions are given in *Protocol 1*.

Protocol 1. Maintenance of *T. monococcum* suspension cultures and preparation of protoplasts

A. *Maintenance of T. monococcum (TM) cultures*

1. Maintain TM cultures in 50 ml P10 medium (11) in 100 ml Ehrlenmeyer flasks, at 25°C at 60 r.p.m. on an orbital platform.

2. Subculture the TM cultures weekly by drawing off 25 ml suspension using wide-bore tissue culture pipettes, and replace with 25 ml of P10 medium.

B. *Isolation of protoplasts*

Isolate protoplasts from suspensions 4–7 days after subculturing.

Protocol 1. *Continued*

1. Draw off 5 ml of TM cell suspension using a sterile Pasteur pipette and transfer to a sterile 9 cm Petri dish.

2. Add 15 ml of protoplast enzyme solution:
   0.05% (w/v) driselase (Sigma)
   0.05% (w/v) cellulase (Onuzua)
   0.05% (w/v) hemicellulase (Sigma)
   0.025% (w/v) pectolyase Y23 (Seishin)
   0.1% bovine serum albumin (fraction V; Sigma)
   3 mM methyl ethyl sulphonate (MES) pH 5.6.
   Spin the enzyme mixture in a bench top centrifuge at maximum speed and sterilize the supernatant by filtration through a nitrocellulose membrane (0.22 μ pore size, Millipore).

3. Seal dishes with parafilm and incubate in the dark for 16 h at 25°C.

4. Agitate the protoplasts gently and pass through successive metal gauze filters of 250 μm, 103 μm, and 50 μm.

5. Add 20 ml solution W (0.7 M mannitol, 10 mM $CaCl_2$ pH 5.6) and spin at 145 g for 5 min. Discard the supernatant.

6. Resuspend the pellet (very gently) in 25 ml solution W and spin at 145 g for 5 min. Discard the supernatant.

7. Resuspend the pellet in solution W to give batches of 4×10^6 protoplasts. The protoplasts are now ready for inoculation (see *Protocol 2*).

We recommend that for the routine subculturing of *T. monococcum* suspension cultures, more than one person maintains sets of flasks, preferably in more than one location so as to minimize the risk of loss of cell lines for technical reasons. As an additional insurance, callus of *T. monococcum* can be maintained on plates containing culture medium solidified with 0.7% (w/v) agarose (Sea Plaque, FMC) and incubated at 25°C in the dark. Maintenance of suspension cultures, preparation of protoplasts and their subsequent culture must be carried out under normal aseptic conditions for plant tissue culture.

3.2 Inoculation of *T. monococcum* protoplasts with plasmid DNA

The method for the polyethylene glycol (PEG) mediated inoculation of plasmid DNAs into *T. monococcum* protoplasts is given in *Protocol 2*.

Protocol 2. Inoculation of protoplasts with plasmid DNA and subsequent culturing

1. Spin each required batch of protoplasts at 145 g for 5 min.

Protocol 2. *Continued*

2. Discard the supernatant and add 0.1 ml of DNA solution containing 50–100 μg of CsCl purified plasmid DNA (sterilized by ethanol precipitation) in 0.05 ml sterile distilled water, 0.05 ml of solution W.[a]

3. Agitate the protoplasts gently and add 0.5 ml of PEG mix, containing 30% (w/v) polyethylene glycol (PEG mol. wt. 6000), 3 mM $CaCl_2$, and 0.7 M mannitol.

4. Agitate the protoplasts gently for 15 sec, then add 5 ml solution W.

5. Incubate at room temperature for 20–30 min.

6. Increase the volume to 50 ml with solution W.

7. Spin at 145 g for 5 min. Discard the supernatant.

8. Repeat steps 6 and 7 with 20 ml solution W each time.

9. After discarding the supernatant, resuspend the protoplasts in 16 ml C8IV medium (12) at a density of 2.5×10^5 protoplasts/ml.

10. Divide the protoplast suspension into 4 ml batches in 5 cm Petri dishes. Prior to use line the bottom of the Petri dishes with 2 ml C8IV medium solidified in 0.7% (w/v) agar.

11. Seal the dishes with parafilm and incubate at 25°C in the dark.

[a] See *Protocol 1* B, step 6.

For routine testing of expression of replicating WDV DNA, incubation periods up to seven days post-inoculation are usually sufficient; after this period protoplast viability declines significantly.

3.3 Extraction and analysis of protoplast DNA

Protocol 3 provides a procedure for the preparation of nucleic acids from protoplasts. The DNA can be digested with restriction endonucleases or with other diagnostic nucleases (e.g. Nuclease S1).

Protocol 3. Isolation of nucleic acids from *T. monococcum* protoplasts

1. Harvest protoplasts by centrifugation (145 g, 5 min). Discard the supernatant.

2. Transfer the protoplasts to a 1.5 ml microcentrifuge tube (e.g. Eppendorf) and spin at *c.* 6500 g for 15 sec. Store samples at −20°C until required.

3. Thaw the protoplast pellet into 0.1 ml Kirby salts (1% (w/v) triisopropyl-naphthalene sulphonide (TNS), 6% (w/v) 4-amino salicylate, 50 mM Tris–HCl pH 8.3, 6% (v/v) buffer equilibrated phenol).

Protocol 3. *Continued*

4. Grind the protoplasts with acid-washed fine sand using a glass rod with a ground-glass coned end. Incubate for 30 min at room temperature.

5. Add 0.15 ml of TE (10 mM Tris–HCl pH 8.3, 0.1 mM EDTA) and vortex for 30 sec.

6. Add 0.4 ml buffer saturated phenol:chloroform (1:1 v/v). Vortex. **Caution: wear surgical gloves and safety glasses. Carry out this step in a fume cupboard.**

7. Spin at 12 000 *g* for 3 min using a bench top microcentrifuge.

8. Transfer the upper, aqueous phase to fresh tubes and re-spin as in step 7, but for 6 min. This step pellets a characteristic sticky residue which is difficult to avoid collecting at the first phenol extraction.

9. Repeat the phenol:chloroform extractions as in steps 6 and 7 a further two times.

10. Add 0.1 vol. 3 M sodium acetate and 2.5 vol. ethanol to precipitate the nucleic acid. Incubate 30 min at −70°C.

11. Spin in a microcentrifuge at 12 000 *g* for 10 min. Discard the supernatant.

12. Wash the pellet in 70% (v/v) ethanol and dry in a vacuum desiccator.

13. Redissolve the pellet in 0.1 ml TE buffer (10 mM Tris–HCl pH 8, 0.1 mM EDTA).

Typically, 5–10 μl (out of 100 μl) of the DNA sample is analysed by electrophoresis through agarose gels, Southern blotting, etc. which are carried out using standard protocols (8). After hybridization with a high specific activity, random-primed ^{32}P labelled probe (9; Chapter 10), the blots are washed (4 × 15 min) in 0.1 × SSC, 0.5% (w/v) SDS at 65°C. Replicating forms of WDV DNA (*Figure 2*) can be readily detected on an autoradiogram after 24 h exposure using an intensifier screen.

3.4 Results

One of our early experiments, using the plasmid pWDVK10D of the wild-type WDV-CJI genome in pIC20R (3), a tandem dimer illustrates a typical response. Following the procedures outlined in *Protocols 1* to *3*, the inoculation of pWDVK10D DNA into protoplasts of *T. monococcum* leads to the production of novel WDV specific DNA forms. These show the same electrophoretic mobilities as the double-stranded forms of the viral DNA found in WDV infected plants, and are first detectable 24 h after inoculation (*Figure 2*), increasing in abundance during the course of the experiment (*Figures 2* and *3*). Restriction enzyme digestions of these WDV-specific forms can be used to confirm that they are dsDNA forms (open-circular, linear, and supercoiled respectively) of the circular

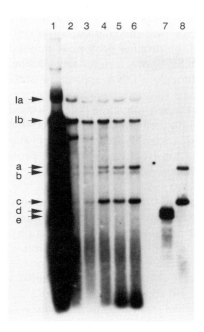

Figure 2. Autoradiogram of a Southern blot, probed with a full length WDV-specific probe, of DNA samples extracted from inoculated protoplasts at 0 days (1), 1 d (2), 2 d (3), 3 d (4), 4 d (5), and 5 d (6) post-inoculation (d.p.i.) with pWDVK10D. Lane 7 is viral ssDNA and lane 8 is dsDNA extracted from infected plants. Bands indicated are inoculum DNA (1a and 1b), progeny dsDNA; open circular (a), linear (b), supercoiled (c); and viral ssDNA (d, e, and lane 7 only).

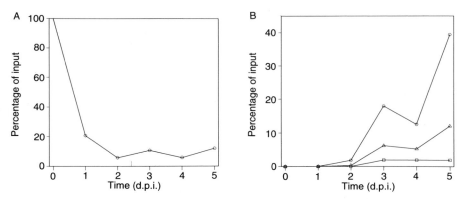

Figure 3. Graphical representation of a densitometric analysis of the autoradiograph shown in *Figure 2*. (A) shows the changes in the amount of inoculum DNA over the course of the experiment. (B) shows the changes in amounts of the progeny DNA forms (open-circular △, linear □, and supercoiled ○). Results are expressed as the percentage of the input DNA amount at $t = 0$.

genome. Inoculated protoplasts start to divide about three days after inoculation. By five to six days post-inoculation, sustained cell division reaches a maximum of *c*. 10% of the original protoplast population. After eight days in culture, protoplast viability shows a significant decline.

Degradation of the inoculum DNA appears to occur immediately following inoculation giving rise to the heterogeneous smear on gel electrophoresis of WDV-specific DNA in the zero-day time point. However, approximately 5% of the input DNA is maintained apparently intact in subsequent time points (*Figures 2* and *3*). This probably represents the amount of DNA which is protected from degradation by delivery into the protoplasts and from which WDV-specific forms are derived.

Additional data is provided by Woolston *et al.* (3) demonstrating that the novel dsWDV-specific DNA forms are the products of replication in the protoplasts and not the accumulated products of homologous recombination of the input DNA. There is no evidence for the replication of the inoculated DNA in the protoplasts and recent experiments in our laboratory suggest that the failure of such a construct to replicate is due to sequences on the plasmid pIC20R, rather than the size of the insert DNA *per se* (H. Reynolds, E. Dekker; unpublished observations). WDV-based vectors, such as pWDV2 (*Figure 1*) produce the same circular dsDNA forms when replicating in *T. monococcum* protoplasts as the wild type genome, except that integrity of any deletion or insertion is retained (3).

4. Transient assay of a reporter gene in plant tissues

4.1 A leaf based assay

The insertion of the chloramphenicol acetyl transferase (*cat*) gene as a 770 bp *Taq* I fragment from pBR328, into a precursor of pWDV2, creates a fusion of the *cat* gene to the virion sense (V1/V2) promoter. This plasmid, termed pWDV-CAT, can replicate in *T. monococcum* protoplasts and from 3–7 days post-inoculation chloramphenicol acetyl transferase (CAT) activity can readily be detected in protoplast extracts.

pWDV-CAT can also be inoculated on to leaves of young wheat plants, by mildly abrading the leaf surface with fine grade carborundum. The methods of inoculation, extraction, and assay of CAT activity are given in *Protocol 4*. CAT activity can be detected in some (but not all) inoculated leaves (*Fig. 4*).

Protocol 4. Leaf inoculation procedure and CAT assay of crude extracts

A. *Leaf inoculation procedure*

1. Grow wheat (*Triticum aestivum* cv Norman) for 14–21 days in peat-based compost, at 15°C, at a photon flux density of 600 $\mu E/m^2/sec$, and a 16 h photoperiod. The plants should be at the 4–5 leaf stage.

Protocol 4. *Continued*

2. Use the youngest fully-expanded leaf for inoculation. Lightly abrade both surfaces of the leaf with fine carborundum using a gloved finger.

3. Spread 20 μl of plasmid DNA (1 mg/ml in sterile distilled water) over approximately 6 cm of leaf length.

4. Leave the plants for 30 min, lightly mist with a water spray, and place back in the growth room.

5. After 72 h, harvest the inoculated area of the leaf and cut into sections small enough to fit into a 1.5 ml microcentrifuge tube. This step should be carried out on ice.

6. Grind the samples in the tubes using a ground-glass rod, acid-washed fine sand and 0.1 ml of extraction buffer (0.05 M Tris–acetate pH 8, 50 mM potassium acetate, 25% (v/v) glycerol, 1 mM EDTA, 10 mM dithiothreitol, 0.5 mM PMSF (Sigma)).

7. Spin the extract in a microcentrifuge at 12 000 g for 5 min, 4°C.

8. Transfer 0.1 ml of the supernatant to clean tube.

B. *CAT assay*

1. To the 0.1 ml of extract (step 8) add 58 μl 0.25 M Tris–HCl pH 7.8 and 2 μl dichloroacetyl-1-2-[^{14}C] chloramphenicol (sp. act. = 2.2 GBq/mmole, concentration = 3.7 MBq/ml; New England Nuclear code NEC-408).

2. Incubate at 37°C for 2 h.

3. Add 1 ml ethyl acetate to each tube and vortex.

4. Spin the tube (12 000 g, 3 min) to separate the aqueous and organic phases.

5. Transfer the upper, organic phase to a clean tube, pierce the lid, and dry down under vacuum (*c*. 30 min).

6. Redissolve the pellet in 10 μl of ethyl acetate and spot on to a 15 cm^2 silica-gel G thin-layer chromatogram (TLC, Sigma). The origin should be 1 cm from the bottom of the plate.

7. Develop the TLC plate in a chromatography chamber containing chloroform/methanol (95 : 5) for *c*. 40 min, or until the solvent front is $\sim$ 10 cm from the origin.

8. Allow the TLC plate to dry in a fume cupboard for 10 min.

9. Autoradiograph at room temperature. An 18 h exposure usually suffices.

Similar methods have been described for the inoculation of WDV-based constructs into dry zygotic embryos by imbibition of a solution containing the DNA (10). In these examples, the ability of the recombinant WDV-reporter gene

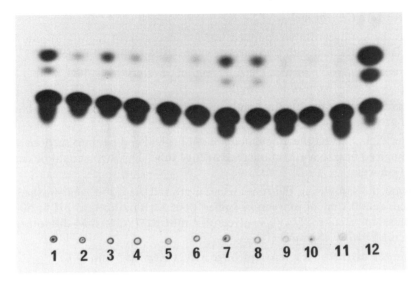

Figure 4. Autoradiograph of a thin-layer chromatogram showing chloramphenicol and its acetylated derivatives which are the substrate and products of the CAT assay respectively. The products of the CAT enzyme are the faster migrating spots. The extracts for the assay were prepared from wheat leaves inoculated separately with pWDV-CAT plasmid DNA (lanes 1–8). Mock inoculated leaves (lanes 9–11) gave no activity. The products from 0.01 units of bacterial CAT enzyme are shown in lane 12.

DNA to replicate in *T. monococcum* correlates with successful detection of reporter gene activity in the tissue inoculation procedures.

4.2 Particle bombardment of barley microspore-derived embryos

We have obtained similar results with pWDV-CAT when inoculated into barley microspore-derived embryos arising from anther culture by microprojectile bombardment techniques. In this system, both replication of novel dsDNA forms and CAT activity can readily be detected seven days post-inoculation. The procedures for barley anther culture and particle bombardment are given in *Procotol 5*.

Protocol 5. Production of barley microspore-derived embryos (anther culture) and their inoculation with DNA using particle bombardment

A. *Anther culture*

1. Sow winter barley (*Hordeum vulgare* cv Igri) in a 2 : 2 : 1 mixture of proprietary compost : perlite : terragreen in seed trays.

Protocol 5. *Continued*

2. Grow in a controlled environment room at 12°C, 16 h photoperiod, 600 μE/m^2 sec, and 80% relative humidity.

3. Spray seedlings at the 2 leaf stage with a systemic fungicide (Dorin, Bayer).

4. At 4 weeks old, vernalize the plants by transferring them to a cold room (4°C, 16 h photoperiod, 160 μE/m^2/sec).

5. Leave for 12 weeks before transferring back to the original growth conditions.

6. After vernalization, transfer the plants to 10 cm pots and feed twice weekly with 100 ml of a 2:1 mixture of Chempak No. 1:Chempak No. 2 (Chempak Ltd).

7. Remove anthers from florets when the microspores are at the mid- to late-uninucleate stage.

8. Place anthers on the medium of Foroughi-Wehr *et al.* (11) except that the major carbon source is replaced with mannitol. Incubate at 25°C. This is called the pretreatment step and replaces cold treatments (12).

9. After 4–6 days, place the anthers on the same medium but containing maltose as the major carbon source. Continue to incubate at 25°C for 2–5 weeks to obtain responding anthers.

B. *Particle bombardment*

1. Transfer developing microspore-derived embryos from responding anthers to a hydrophobic-edged nitrocellulose membrane (0.45 μm pore size; Sartorius) in a 5 cm Petri dish containing 1 ml of liquid culture medium. Up to 5 anthers can be used per filter, depending on the response.

2. Perform particle-gun bombardment with a Biolistic Gene Gun[a] (DuPont); detailed instructions accompany the gun and can be found in ref. 13.

3. Coat DNA on to tungsten particles (median diameter 1.1 μm: 0.3 mg tungsten coated with 0.5 μg DNA per shot.)

4. Fire DNA-coated tungsten particles at the tissue using a 0.22 calibre long blank (available from any gun shop). Position the plates containing the embryos 7.5 cm from the macroprojectile stopper plate.

5. After shooting, return the embryos to the same incubation regime.

[a] Alternative systems may be used.

5. Future refinements

The techniques described in this chapter have been developed using wild-type WDV DNA or virus genomes containing simple additions to a V2 ORF deletion mutant, e.g. as in pWDV-CAT. Currently, several WDV replicon based plasmids are being developed which should allow the convenient insertion and assay of any transcriptional regulatory sequences fused to a reporter gene. In addition, methodologies are being developed to allow the quantification of promoter activity by introducing a reference gene to act as an internal standard. It will be of interest to determine the behaviour of regulated promoters when present at high copy number on a replicating vector. In addition, it should be possible to extend the range of tissues which can be inoculated by this procedure.

Acknowledgements

I wish to thank Gary Creissen, Elise Dekker, Helen Reynolds, Nicola Stacey, and Cris Woolston for the development of the methods and plasmids described in this chapter. In addition, thanks are due to GC and HR for the critical reading of the manuscript and Julie Hofer for the design of *Figure 1*.

References

1. Stanley, J. (1985). *Adv. Virus Res.*, **30**, 139.
2. Davies, J. W., Stanley, J., Donson, J., Mullineaux, P. M., and Boulton, M. I. (1987). *J. Cell Sci. Suppl.*, **7**, 95.
3. Woolston, C. J., Reynolds, H. V., Stacey, N. J., and Mullineaux, P. M. (1989). *Nucleic Acids Res.*, **17**, 6029.
4. Schalk, H. J., Matzeit, V., Schiller, B., Schell, J., and Gronenborn, B. (1989). *EMBO J.*, **8**, 359.
5. Wahl, G. (1989). *Strategies*, **2**, 17.
6. Boyer, H. W. and Roulland-Dussoix, D. (1969). *J. Mol. Biol.* **41**, 459.
7. Lorz, H., Baker, B., and Schell, J. (1985). *Mol. Gen. Genet.* **199**, 178.
8. Maniatis, T., Fritsch, E. F., and Sambrook, J. (1982). *Molecular cloning: a laboratory manual*. Cold Spring Harbor Laboratory Press, New York.
9. Feinberg, A. P. and Vogelstein, B. (1984). *Anal. Biochem.*, **132**, 641.
10. Topfer, R., Gronenborn, B., Schell, J., and Steinbiss, H. H. (1989). *Plant Cell*, **1**, 133.
11. Foroughi-Wehr, B., Mix, G., Gaul, H., and Wilson, H. M. (1976). *Z. Pflanzenzucht.*, **77**, 198.
12. Roberts-Oehschlager, S. L. and Dunwell, J. M. (1991). *Plant Cell Rep.* **9**, 631.
13. Sanford, J. C. (1988). *Trends in Biotechnology*, **6**, 299.

3

RNA-based viruses

PHILIP KRONER and PAUL AHLQUIST

1. Introduction

This chapter describes the use of protoplasts to study various aspects of the replication cycle of (+) strand RNA viruses that infect plants. The genetic analysis of these viruses, which comprise the overwhelming majority of plant viruses, has been accelerated in recent years by the development of methods that produce infectious viral RNA from genetically manipulatable cDNA clones. Using this technology, these viruses can be studied both *in vivo* and *in vitro* in a variety of experimental systems using, for example whole plants, protoplasts, and extracts purified from infected tissue. Each of these systems has its own advantages, such as emphasizing aspects of the virus lifecycle. Many important issues relating to the movement of virus infection, for example, require investigation at the level of whole plants or tissues. For many studies, however, protoplasts provide an especially valuable system. Protoplasts are relatively easy to isolate and can be inoculated with high efficiency with viral RNA. In addition, inoculation of protoplasts results in a synchronized infection that allows the study of all intracellular processes involved in virus production including RNA uncoating and the initiation of infection, all aspects of RNA replication, and encapsidation of viral RNA. Because most protoplasts show a general permissiveness for the replication of many virus types, they provide a useful system for analysis of the interactions between viruses with a divergent host range. Work in our own laboratory has concentrated primarily on the molecular biology of the bromoviruses, brome mosaic virus (BMV) and cowpea chlorotic mottle virus (CCMV), and here we describe some tested procedures that have proven useful in applying barley protoplasts to these investigations (1–3). Specifically, we have chosen to focus on methods for the synthesis of expressible viral cDNA, the *in vitro* production of infectious transcripts, the isolation and inoculation of barley protoplasts, and the analysis of viral replication in infected cells. Many of these procedures should be directly applicable, and most others easily adaptable, to the study of other plant viruses. Similar studies have been conducted, for example, with tobacco mosaic virus in tobacco protoplasts (4) and with cowpea mosaic virus in cowpea protoplasts (5). Additional examples can be found in the review by Motoyoshi (6).

2. Synthesis and convenient fusion of viral cDNA to suitable promoters for *in vitro* transcription

The combination of cDNA cloning methods and powerful *in vitro* transcription systems has contributed to major advances in the analysis of RNA viruses. Firstly, sequencing RNA virus genomes through cDNA clones has revealed the underlying unity of their replication mechanisms, despite extreme differences in morphology and genetic organization (7). Secondly, fusion of viral cDNA to efficient RNA polymerase promoters has allowed the *in vitro* synthesis of infectious RNA and permits the full range of recombinant DNA techniques to be used in the genetic analysis of these viruses.

When constructing expressible cDNA clones, the correct fusion of the RNA polymerase promoter to the viral cDNA is often an important factor for the *in vitro* production of infectious transcripts. Studies with BMV and other viruses have shown that, among other factors, incorrect promoter fusions resulting in the addition of one or more non-viral bases to the 5' end of viral transcripts can substantially reduce infectivity *in vivo*. Since many current methods for viral cDNA/promoter fusion result in the addition of non-viral bases to the 5' end, time-consuming mutagenesis procedures may be required to restore the wild type viral sequence. We present here a simple cDNA cloning strategy that correctly fuses the viral sequence to the selected promoter during cDNA synthesis. The basis of this approach is illustrated in *Figure 1* for the case of fusing BMV RNA1 cDNA to a bacteriophage T7 promoter. First-strand cDNA synthesis is performed as described by Allison *et al.* (8) so that a unique restriction site is added to the 3' end of the viral sequence when cDNA synthesis is completed. Correct fusion of the RNA polymerase promoter occurs during second-strand cDNA synthesis by using a specialized primer that consists of three parts (5' to 3'); first, the sequence of a unique restriction site to enhance cloning; second, the 18 base sequence that constitutes the fully-active bacteriophage T7 RNA polymer-

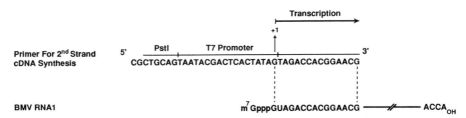

Figure 1. Example of a primer for second-strand cDNA synthesis designed to provide correct fusion of viral sequences to a T7 RNA polymerase promoter. The example shown is for cDNA cloning of BMV RNA1. The primer consists of three segments: a built-in restriction site to provide a cohesive end to facilitate ligation of the cDNA into a cloning vector; the T7 promoter sequence (9); and the actual primer sequence that anneals to the viral cDNA after first-strand synthesis. The first transcribed base (designated +1) is also the last base of the T7 RNA polymerase promoter.

ase promoter as defined by Milligan *et al.* (9); and third, the 5' sequence of the viral RNA being cloned. The presence of unique restriction sites in the primers used for first- and second-strand synthesis facilitates cloning of the double-stranded cDNA and improves the recovery of full-length clones. Because the T7 RNA polymerase promoter is included in the primer for second-strand synthesis, the full-length cDNA can be inserted into any convenient plasmid vector.

3. *In vitro* synthesis of infectious transcripts

Infectious BMV transcripts are efficiently produced from constructs containing the T7 bacteriophage RNA polymerase promoter by following the general conditions of Melton *et al.* (10). The reaction conditions are adjusted to facilitate the incorporation of a cap (m^7GpppG) at the 5' end of the transcripts (11). Capping transcripts is important for infectivity, as BMV transcripts lacking a cap are not detectably infectious to whole plants and only weakly infectious to protoplasts. The conditions given in *Protocol 1* result in the capping of approximately 60% of the transcripts and increase the infectivity of BMV transcripts to protoplasts by at least 10-fold. The inclusion of m^7GpppG in the transcription reaction increases the infectivity of some viral transcripts even if the wild type viral RNA does not normally contain a cap structure at the 5' end. For example, the addition of a cap to cowpea mosaic virus transcripts *in vitro* increases infectivity to protoplasts 2- to 3-fold (5), even though the 5' end of the wild type genome is not capped, but is instead linked to a small protein, VPg. For this virus, and with BMV, the cap structure may increase infectivity by enhancing the stability and/or translatability of the viral RNA.

Since substitution of m^7GpppG with the less expensive GpppG reduces infectivity of the transcripts by only half, we routinely use the latter in our transcription reactions. At the completion of the transcription reaction, DNAaseI is used to digest DNA templates since their presence causes inefficient inoculation of barley protoplasts. The integrity and yield of RNA that is produced can be monitored by *in vitro* translation and/or agarose gel electrophoresis. Similar reaction conditions in the absence of cap also produce radioactive RNA for Northern blot analyses as noted in *Protocol 3*. For synthesis of transcripts, and in all subsequent manipulations involving RNA, extreme care should be taken to prevent RNA degradation by contaminating ribonucleases. To minimize such contamination autoclave or bake glassware, purchase RNAase-free reagents when possible, and if necessary, treat solutions with diethylpyrocarbonate (12) prior to sterilization.

Protocol 1. Procedure for *in vitro* synthesis of infectious transcripts

1. For a 100 μl reaction containing 2 μg of DNA template, combine at room temperature:

- 20 μl of 5 × transcription buffer [a]

Protocol 1. *Continued*

- 10 μl of 100 mM DTT
- 10 μl of nucleotide mix[b]
- 25 μl of 2 mM GpppG
- 32 μl of water
- 2 μl of linearized DNA template (1 mg/μl)
- 1 μl of T7 RNA polymerase (20 units/μl)

Mix well and incubate at 37°C for 60 min.

2. Add 2 μl of DNAaseI (1 unit/μl, Promega) and incubate at 37°C for 15 min.
3. Extract the reaction mixture twice with an equal volume of phenol/chloro-form/isoamyl alcohol (24:24:1) (15).
4. Precipitate RNA from the aqueous phase by adding 0.1 vol. 3 M sodium acetate (pH 6.0) and 0.65 vol. isopropanol. Store at −20°C overnight or at −80°C until frozen (about 30 min).
5. Pellet RNA by centrifugation (12 000 g, 10 min). Discard the supernatant. Wash the pellet with 1 ml of 70% EtOH, repeat the centrifugation and discard the supernatant. Dry the RNA pellet under vacuum.
6. Resuspend the RNA in 10 μl of 10 mM Tris–HCl, 0.1 mM EDTA, pH 7.5.

[a] 200 mM Tris–HCl pH 7.5, 30 mM $MgCl_2$, 10 mM spermidine, 50 mM NaCl.
[b] 4 mM each ATP, CTP, TTP; 0.75 mM GTP in 5 mM Tris–HCl, 0.1 mM EDTA, pH 8.0.

4. Isolation of barley protoplasts

Many factors influence the isolation of viable protoplasts, and experimental protocols must often be modified for specific tissues, organs, or cell cultures (13). The procedure for isolation of barley mesophyll protoplasts described in *Protocol 2* is, essentially, that developed by Loesch-Fries and colleagues (14, 15) with modifications made by R. French and others in our laboratory. For consistent results, maintain uniform plant growth conditions as the physiological state of the plants is an important factor influencing protoplast yield and efficiency of infection. Barley grows well in peat moss and vermiculite (1 : 3) watered with half-strength Hoaglands solution once each day. Protoplasts are isolated from 6-day-old seedlings (6 to 8 cm long leaves) grown at 24°C in a growth chamber with lights (100 $\mu Em^{-2} sec^{-1}$) set at a 16 h day-length. This procedure works well with many varieties of barley; Morex, Robust, and Dickson have been used with equal success. Protoplast isolation and inoculation are performed in a laminar-flow hood using aseptic technique.

Protocol 2. Procedure for isolation of barley protoplasts

1. Sterilize a glass plate, 4 single-edged razor blades, and a fine mesh polyester screen (approximately 0.02 inch square openings) with 70% EtOH and allow to dry.

2. Dispense 25 ml of freshly made enzyme solution[a] into two Petri dishes.

3. Harvest 4 g barley leaves using an additional razor blade. Gently cut 4 to 5 leaves at a time at the soil surface. It is not necessary to wash or surface sterilize leaves. Divide the leaves into 2 g aliquots.

4. Gently hold one aliquot of leaves vertically and tap on the glass plate to align the cut ends. Lay the leaves flat and, beginning with the tips, cut the leaves transversely with a sterile razor blade into slices 1 to 2 mm wide. Be careful not to crush the leaves. Use two razor blades for each aliquot of leaves. Discard the last 1 to 2 cm of stem.

5. Use a razor blade to transfer the tissue slices to one of the Petri dishes and gently disperse and wet the tissue sections in the enzyme solution.

6. Starting with new razor blades, cut the second aliquot of leaves. Transfer the tissue sections to the second Petri dish and place both Petri dishes in a 30°C incubator. Maintain in the dark for 3 h.

7. Remove the Petri dishes from the incubator and gently swirl for about 30 sec to dislodge protoplasts from partially digested tissue.

8. Pour the solution through the polyester screen into a 100 ml beaker.

9. Transfer the filtrate containing the protoplasts to four 16 × 100 mm glass tubes.

10. Sediment tissue debris and cells by centrifugation (50 g, 2.5 min). Remove the supernatant with a pipette and discard. Be careful not to disturb the soft, dark green pellet.

11. Slowly add 1 ml of 10% (w/v) mannitol (sterilized by autoclaving) to each tube. Hold the tubes at a shallow angle and resuspend the pellets using a gentle rocking motion.

12. Add a further 8 ml of 10% (w/v) mannitol to each tube. Add the first 4 ml down the side of the test-tube and the last 4 ml directly into the protoplast suspension so that the protoplasts are evenly distributed.

13. Fill a 9 inch Pasteur pipette with about 2 ml of 20% (w/v) sucrose (sterilized by autoclaving). Place the tip of the pipette at the bottom of a tube containing the protoplast suspension, then allow the sucrose solution to slowly fill the bottom of the tube. There should be a distinct line at the interface of the sucrose solution and the protoplast suspension.

14. Separate the protoplasts from the cell debris by centrifugation (50 g, 8.5 min).

Protocol 2. *Continued*

15. Collect protoplasts from the top of the sucrose pad using a Pasteur pipette and transfer to a beaker containing 20 ml of 10% (w/v) mannitol. The tissue debris will form a pellet in the bottom of the tube.

16. Count protoplasts in a 0.9 μl sample of the final suspension using a haemacytometer. Average yields are 2 to 3×10^5 protoplasts per ml.

---

[a] Combine 5 g mannitol, 1 g Cellulysin cellulase (Calibiochem), 50 mg Macerozyme R-10 (Yakult Honsha Co.), 50 mg BSA (Sigma), and 40 ml of water. Stir 10 min, then adjust to pH 9.0 with 1 M KOH. The solution should clear immediately. Adjust to pH 5.7 with 0.3 M citric acid, make up to 50 ml with water and filter through a 0.45 μm filter.

---

5. Inoculation of barley protoplasts

We routinely introduce viral RNA into protoplasts in the presence of PEG and $CaCl_2$ as described in *Protocol 3*. The method is simple, results in the consistent infection of a high percentage of the protoplasts, and can easily accommodate a large number of samples. For experiments with BMV we routinely inoculate 10^5 protoplasts with the RNA produced from 0.5 μg of each DNA template (about 5 μg of each genomic RNA). This produces a level of infection that is equivalent to that seen in protoplasts inoculated with about 1 μg of virion RNA. We have also investigated electroporation as an alternative method for introducing viral RNA into protoplasts. Under optimal conditions the level of viral replication in protoplasts inoculated by electroporation is nearly equivalent to that attained by the $PEG/CaCl_2$ method. Electroporation (optimal conditions are 300 V, 200 μF, 10 ms; 4 mm gapped stainless steel plate electrodes) was performed on ice with a Model 450 Electroporation System (Promega) with 10^5 cells suspended in 1.0 ml of 0.45 M mannitol, 70 mM KCl, and 5 mM MES (pH 5.7). After electroporation, samples were allowed to recover on ice for 30 min, and then processed as described in *Protocol 3* steps 7 to 10.

Protocol 3. Procedure for inoculating barley protoplasts

Reagents

- PEG inoculation buffer (40% w/v PEG, 3 mM $CaCl_2$): the source of the polyethylene glycol is important. PEG 1540, pharmaceutical grade from Polysciences, Inc. works well. Dissolve 20 g PEG in 25 ml of water. Add 1.5 ml of 0.1 M $CaCl_2$, make to 50 ml with water, and filter sterilize. Store in the dark at 4°C. This solution is stable for at least six months.

- 10 × Aoki salts (19):

2 mM KH_2PO_4	0.27 g
10 mM KNO_3	1.01 g
10 mM $MgSO_4$ (heptahydrate)	2.46 g

Protocol 3. *Continued*

10 μM KI	1 ml of 0.16% (w/v) solution
1 μM CuSO$_4$	0.1 ml of 0.25% (w/v) solution
100 mM CaCl$_2$	14.7 g

Add the first five ingredients to 900 ml of water and stir until they are all dissolved, then add the CaCl$_2$. When the CaCl$_2$ is dissolved add water to 1 litre and autoclave.

- Protoplast medium: Dissolve 10 g mannitol in 70 ml of water, then add 10 ml of 10 × Aoki salts and 100 μl of 10 mg/ml gentamicin sulfate (Sigma). Adjust the pH to 6.5 with 0.1 M KOH, make up to 100 ml with water and filter-sterilize. Store at room temperature.

Inoculation procedure

1. For each inoculation aliquot 10^5 protoplasts into a 12 × 75 mm glass test-tube.

2. Pellet the cells by centrifugation (50 g, 2 min). Remove the supernatant with a Pasteur pipette and discard. Be careful not to disturb the soft cell pellet.

3. Suspend the cells in 1 drop (about 30 μl) of 10% (w/v) mannitol.

4. Add viral RNA (in 15 μl or less of 10 mM Tris–HCl, 0.1 mM EDTA pH 7.5) to the cell suspension and immediately add 100 μl of PEG inoculation buffer. Mix well by gently tapping the tube for 10 sec.

5. Rapidly add 1 ml of 10% (w/v) mannitol. Mix by tapping the tube for 5 sec. Place the tube on ice and proceed to the next inoculation.

6. After the final inoculation hold the tubes on ice for 15 min.

7. Pellet the cells by centrifugation (50 g, 3 min). Discard the supernatant.

8. Gently resuspend the cells in 1 ml of 10% (w/v) mannitol.

9. Pellet the cells by centrifugation (50g, 2 min) and discard the supernatant.

10. Resuspend the cells in 0.5 ml of protoplast medium and transfer to a 1.5 ml microfuge tube or a multi-well tissue culture dish. If a tissue culture dish is used, seal the edges with tape to avoid dehydration. Maintain at 24°C with constant illumination (25 μEm^{-2} sec^{-1}).

6. Analysis of viral replication in protoplasts

6.1 Isolation of virions from infected protoplasts

The specific interactions between the viral RNA and the coat protein that lead to encapsidation can be investigated in protoplasts when this technique is used in combination with a mutational analysis of the viral genome. The technique described in *Protocol 4* results in the isolation of whole virions from infected protoplasts. The most critical step in this procedure is the method used to break open cells since this requires that the protoplasts be broken open without damaging the virions. The most consistent results are obtained when protoplasts are held on ice and broken open by sonication. Optimal sonication conditions will depend on the particular probe and power unit used. Following the procedures in *Protocol 4*, 50 to 100% of the total viral RNA present in protoplasts infected with either BMV or CCMV RNA can be isolated (*Protocol 5*) as encapsidated RNA.

Protocol 4. Isolation of virions from infected protoplasts

1. Inoculate 10^5 protoplasts with viral transcripts as described in *Protocol 3*. Resuspend protoplasts in 1.0 ml of protoplast medium and aliquot 500 μl into two 1.5 ml microcentrifuge tubes. Following incubation for 24 to 48 h, isolate total RNA from one sample according to the method described in *Protocol 5*. The viral RNA in this sample represents the total virion RNA present in the protoplasts (i.e. encapsidated plus unencapsidated RNA).

2. Add 100 μl of 5× virion buffer (2.5 M sodium acetate, 0.4 M magnesium acetate, pH 4.8) to the second sample and chill on ice prior to sonication.

3. Sonicate the protoplasts for 30 sec, e.g. at a setting of −041 (20 watts) using a Braun-Sonic 2000 sonicator fitted with a microprobe. Keep the protoplast suspension on ice during and after sonication.

4. Rinse the microprobe thoroughly first with 95% ethanol, then with 1 × virion buffer before sonicating the next sample. Use fresh washing solutions after each sonication.

5. After the final sonication, pellet cell debris by centrifugation at 12 000 g for 3 min. Transfer the supernatant to a fresh microcentrifuge tube.

6. Add approximately 50 μg of a carrier virus (e.g. tobacco mosaic virus) and 160 μl of 30% (w/v) polyethylene glycol 8000 (Sigma) to the supernatant and allow the virions to precipitate on ice for 30 min. Addition of heterologous carrier virus improves the efficiency of precipitation of low amounts of virus and will not interfere with the Northern analysis later.

7. Pellet the virions by centrifugation at 12 000 g for 10 min, and discard the supernatant. Collect the residual supernatant by centrifugation (12 000 g, 2 min) and discard. Resuspend the pellet in 500 μl of 1 × virion buffer.

Protocol 4. *Continued*

8. Digest any unencapsidated RNA that may have coprecipitated with the virions by adding 160 μl of 0.1 M CaCl$_2$ and 300 units of micrococcal nuclease (Boehringer-Mannheim). Incubate at 37°C for 15 min.

9. Add 20 μl of 0.1 M EGTA to inhibit the nuclease activity and then isolate the RNA following the procedure described in *Protocol 5*. The intact viral RNA in this sample represents encapsidated RNA.

10. Determine the relative proportion of viral RNA in protoplasts that is encapsidated by comparing the signals obtained from Northern analysis of equivalent portions of the RNA obtained at steps 1 and 9.

Protocol 5. Isolation of nucleic acids from infected protoplasts

1. Add 200 μl of freshly made extraction buffer[a] directly to the 0.5 ml of incubation medium containing the protoplasts, then vortex briefly (10 sec) at high speed to disrupt cells.

2. Extract the lysed cell mixture with 500 μl of phenol/chloroform/isoamyl alcohol (24:24:1) (12).

3. Separate the phases by centrifugation at 5°C (12 000 g, 5 min).

4. Transfer the upper aqueous phase to a fresh tube and repeat the phenol extraction.

5. Transfer the aqueous phase to a fresh tube containing 65 μl of 3.0 M sodium acetate (pH 6.0). Add 500 μl of isopropanol and place the tubes at -70°C for 15 to 30 min or at -20°C for overnight.

6. Pellet the nucleic acids by centrifugation at 5°C (12 000 g, 10 min). Discard the supernatant. Wash the pellet with 1 ml of 70% ethanol and recover the nucleic acids by centrifugation at 5°C (12 000 g, 10 min).

7. Discard the supernatant and dry the pellet under a vacuum.

8. Resuspend the pellet in 30 μl of water.

[a] Extraction buffer is 2.0 ml of 0.5 M glycine, 0.5 M NaCl, 5 mM EDTA pH 9.5, plus 0.5 ml 20% SDS (w/v), and 0.5 ml bentonite (see *Protocol 6*).

Protocol 6. Preparation of bentonite

1. Suspend 100 g of bentonite (325 mesh, E. H. Sargent and Co.) in 1.5 litres of water and stir overnight at room temperature.

2. Sediment large particles by centrifugation (160 g, 15 min). Transfer the supernatant to fresh bottles and collect the smaller particles by centrifugation (5800 g, 30 min). Discard the supernatant.

Protocol 6. *Continued*

3. Resuspend the pellet in 1.5 litres of water and repeat the centrifugation cycle in step 2, saving the pellet from the second centrifugation.

4. Suspend the pellet in 1.5 litres of 0.1 M EDTA (pH 7.0) and stir overnight at room temperature.

5. Repeat the centrifugation cycle in step 2 and resuspend the pellet in TBE buffer (12) at a final concentration of approximately 50 mg/ml. The dry weight of bentonite is roughly 10% of the weight of the pellet from the final centrifugation. Store at 5°C.

6.2. Northern blot analysis of viral RNA accumulation

After isolating the nucleic acids from infected protoplasts, the level of viral RNA is determined by standard Northern blotting procedures. For the most consistent detection of positive-strand and negative-strand RNA, and for accurate size determination, nucleic acids should be denatured prior to electrophoresis. We follow the glyoxal denaturation method of McMaster and Carmichael (17). Briefly, the RNA from about 10 000 inoculated protoplasts is incubated in a solution containing 1.0 M deionized glyoxal (Aldrich), 50% dimethyl sulphoxide (Me$_2$SO), 10 mM sodium phosphate pH 7.0, in a final volume of 15 μl at 50°C for 1 h. The reaction mixture is cooled on ice and 2 μl of 50% glycerol, 10 mM sodium phosphate, and 0.2% bromphenol blue is added. The samples are electrophoresed on horizontal 1% agarose gels containing 10 mM sodium phosphate buffer (pH 7.0) with constant buffer recirculation. Positive-strand RNA levels can also be determined by Northern blotting techniques following electrophoresis of the nucleic acids in non-denaturing horizontal 1% agarose gels containing TBE buffer (12). In this case, the integrity and yield of protoplast RNA in each sample can be monitored by staining the gel with EtBr.

Following electrophoresis, native or denatured RNA is transferred to nylon membranes overnight by the capillary blotting method of either Southern (18) or Thomas (19) in the presence of 10 × SSC (1.5 M NaCl, 0.15 M sodium citrate, pH 7.0). The filters are baked at 80°C (with or without a vacuum) for at least 2 h, pre-hybridized for at least 1 h at 60°C in a solution containing 50% formamide, 50 mM sodium phosphate pH 6.5, 0.8 M NaCl, 1 mM EDTA pH 8.0, 10 × Denhardt's solution (12), 0.25 mg/ml sheared, denatured salmon sperm DNA, 0.5 mg/ml yeast RNA, and 0.5% SDS. Hybridization is for 12 to 24 h at 60°C in 15 to 20 ml of the same buffer containing 2 × 10^7 c.p.m. of radioactive RNA (*Protocol 7*). Non-specific hybridization is removed by washing the filters as described by Thomas (19) except that the high-stringency wash is in 0.2 × SSC, 0.2% SDS at 60°C. We find that ^{32}P-labelled viral RNA is the most sensitive probe for detecting BMV RNA on nylon filters (1).

Protocol 7. Procedure for synthesis of ^{32}P-labelled viral RNA probes

1. In a 1.5 ml microcentrifuge tube combine at room temperature:
- 2.5 μl of water
- 5 μl of 5 × transcription buffer (*Protocol 1*)
- 2.5 μl of 0.1 M dithiothreitol
- 2.5 μl of nucleotide mixture (ATP, GTP, TTP at 5 mM; CTP at 100 μM)
- 10 μl of [α-^{32}P]CTP (100 μCi, 3000 Ci/mmol; Amersham)
- 1 μg of linearized DNA template
- 1 μl (10 to 20 units) of the appropriate RNA polymerase

Incubate at 37°C for 60 min.

2. Add 1 unit of RQ DNAaseI (1 unit/μl, Promega) and incubate at 37°C for 15 min.

3. Add 2.5 μl of 0.5 M EDTA (pH 8.0) and 75 μl of STE (10 mM Tris, 1 mM EDTA, 0.2% SDS, 0.5 M NaCl pH 7.5).

4. Remove unincorporated nucleotides from the reaction mixture by following the G50 Sephadex spun column procedure of Maniatis *et al.* (12). Determine the amount of [α-^{32}P]CTP incorporated into RNA by counting an appropriate dilution in a scintillation counter. A typical reaction yields 10^8 c.p.m. of radioactive RNA.

Acknowledgements

We thank Radiya Pacha for permission to describe the protoplast virion isolation procedure and for critical review of the manuscript. Development and testing of the above methods was supported by grants from the National Institutes of Health (GM35072) and the National Science Foundation (DMB-9004385).

References

1. Kroner, P., Richards, D., Traynor, P., and Ahlquist, P. (1989). *J. Virol.*, **63**, 5302.
2. Pacha, R., Allison, R. F., and Ahlquist, P. (1990). *Virology*, **174**, 436.
3. Traynor, P. and Ahlquist, P. (1990). *J. Virol.*, **64**, 69.
4. Ishikawa, M., Meshi, T., Watanabe, Y., and Okada, Y. (1988). *Virology*, **164**, 290.
5. Vos, P., Jaegle, M., Wellink, J., Verver, J., Eggen, R. Van Kammen, A., and Goldbach, R. (1988). *Virology*, **165**, 33.
6. Motoyoshi, F. (1985). In *Plant protoplasts* (ed. L. C. Fowke and F. Constabel), p. 119. CRC, Boca Raton, Florida.
7. Goldbach, R. and Wellink, J. (1988). *Intervirology*, **29**, 260.
8. Allison, R. F., Janda, M., and Ahlquist, P. (1988). *J. Virol.*, **62**, 3581.

9. Milligan, J. F., Groebe, D. R., Witherell, G. W., and Uhlenbeck, O. C. (1987). *Nucleic Acids Res.*, **15**, 8783.
10. Melton, D. W., Kreig, P. A., Rebagliati, M. R., Maniatis, T., Zinn, K., and Green, M. R. (1984). *Nucleic Acids Res.*, **12**, 7035.
11. Nielsen, D. A. and Shapiro, D. J. (1986). *Nucleic Acids Res.*, **14**, 5936.
12. Maniatis, T., Fritsch, E. F. and Sambrook, J. (1982). *Molecular cloning: a laboratory manual*, Cold Spring Harbor Press, New York.
13. Eriksson, T. R. (1985). In *Plant protoplasts* (ed. L. C. Fowke and F. Constabel), p. 1. CRC, Boca Raton, Florida.
14. Loesch-Fries, L. S. and Hall, T. C. (1980). *J. Gen. Virol.*, **47**, 323.
15. Samac, D. A., Nelson, S. E., and Loesch-Fries, L. S. (1983). *Virology*, **131**, 455.
16. Aoki, S. and Takebe, I. (1969). *Virology*, **39**, 439.
17. McMaster, G. K. and Carmichael, G. G. (1980). *Proc. Natl. Acad. Sci. USA*, **74**, 4835.
18. Southern, E. M. (1975). *J. Mol. Biol.*, **98**, 503.
19. Thomas, P. S. (1980). *Proc. Natl. Acad. Sci. USA*, **77**, 5201.

Bacteria

MICHAEL J. DANIELS

1. Diversity and taxonomy

Some of the most serious diseases of crops are caused by bacteria. Although bacterial diseases can be found in all parts of the world, they are particularly important in tropical and subtropical countries where warm, humid conditions are ideal for bacterial growth. The pioneering work of Smith (1) nearly a century ago laid the foundations for our knowledge of plant pathogenic bacteria although these organisms have been somewhat neglected by plant pathologists until recently. With the emergence of 'molecular plant pathology', there has been an upsurge of interest in the major groups of plant pathogenic bacteria which represent excellent systems for detailed study.

True plant pathogenic bacteria, that is those which are capable of parasitizing and causing disease to an otherwise healthy growing plant, belong to a small number of genera. Full details are given in the recent monograph by Bradbury (2), and *Table 1* lists the major genera. The Gram-negative genera *Agrobacterium*, *Erwinia*, *Pseudomonas*, and *Xanthomonas* have received most attention from

Table 1. Major groups of plant pathogenic bacteria

Genus	Gram reaction	Types of disease caused	Comments
Agrobacterium	−	tumours	
Clavibacter	+	wilt, canker, leaf spot, stunt	Formerly *Corynebacterium*
Curtobacterium	+	wilt, leaf and stem spot	Formerly *Corynebacterium*
Erwinia	−	necrosis, soft rot, wilt	
Pseudomonas	−	wilt, leaf spot, canker	*P. syringae* includes about 50 pathovars
Rhodococcus	+	fasciation, galls	Formerly *Corynebacterium*
Spiroplasma	−	stunt, wilt	Mollicutes, leafhopper transmitted
Xanthomonas	−	leaf spot, rot, wilt, canker	*X. campestris* includes more than 130 pathovars

molecular geneticists (3). These organisms can be cultured in simple laboratory media, readily inoculated into plants for pathogenicity tests, and an extensive set of general and special purpose cloning vectors, together with techniques for genetic exchange, permit direct genetic studies of pathogenicity. In the case of *Agrobacterium*, the ability of the organism to transfer T-DNA into plant cells and thereby transform them genetically has stimulated much work on the basic biology of the plant–bacterial interaction (4, 5). The Gram-positive pathogens (*Clavibacter*, *Curtobacterium*, and *Rhodococcus*) are more difficult to handle and still await the development of satisfactory genetic methodology. The term 'fastidious prokaryotes' is often used to describe a diverse group of pathogens which infect xylem or phloem, causing diseases of the 'yellows' group. Until the late 1960s these diseases were believed to be caused by viruses, but results from antibiotic therapy, electron microscopy, and, in some cases, culturing, showed that the pathogens are in fact prokaryotes. Some, which are confined to phloem, lack cell walls and have similarities to animal mycoplasmas (6). The majority of these cannot be cultured and are called 'mycoplasma-like organisms' (MLOs); a subset, the spiroplasmas, are however, cultivatable *in vitro* (7). The xylem-limited bacteria have cell walls and were originally called 'rickettsia-like organisms', although the successful culture and characterization of some members of the group has led to the abandonment of this designation. Apart from the cloning of some genes and the use of cloned DNA for diagnostic probing (8), little molecular genetic work relevant to pathogenicity has been reported.

Although a discussion of bacterial taxonomy is not appropriate in this book, aspects of nomenclature deserve comment because of their relevance to the important question of host specificity (see also Chapter 6). Certain species, e.g. *Pseudomonas syringae* and *Xanthomonas campestris* cause disease on a wide range of plant species belonging to many families, but generally a pathogenic strain isolated from a particular diseased plant can only infect the same or other closely related species (9). The strains from different plant species are in general indistinguishable from one another using standard bacteriological tests, and are designated 'pathovars'. Such pathovars are usually named after the plant host from which they were originally isolated (2). Thus *P. syringae* pathovar (pv) *phaseolicola* infects *Phaseolus* (bean), *X. campestris* pv *citri* infects citrus plants, etc. Molecular genetic studies have confirmed the relatedness of different pathovars of certain species (10, 11). Little is yet known about the genetic features of pathovars that are responsible for host specificity; there is some evidence that avirulence genes may be involved (12). The curious concept of 'avirulence' genes carried by a pathogen comes from classical genetic studies with fungal pathogens which have shown that, in many cases, specificity is determined by matching genes in host and pathogen. The biochemical responses which lead to an incompatible interaction (i.e. no disease) are triggered only if the plant contains a dominant resistance gene and the pathogen also carries a dominant allele of a matching gene. Hence, the product of the latter gene conditions avirulence in the pathogen (13). Many pathovars are further subdivided into races which are defined as strains specific to certain genotypes or cultivars of a single plant species. There are several examples where avirulence genes in bacterial pathogens interacting with resistance genes in the host have been shown to determine race–cultivar specificity (14).

2. Diseases and habitats

The disease symptoms caused by bacteria infecting susceptible host plants fall into a small number of general classes:

(a) *Leaf spots*. Localized lesions, each developing from a separate infection event. The tissue may be 'water-soaked' (i.e. presenting a greasy appearance) or necrotic. Spread of the lesions is limited. Many *P. syringae* and *X. campestris* pathovars cause leaf spots.

(b) *Vascular wilts*. These are caused by bacteria which colonize the xylem and thereby interfere with water transport. *P. solanacearum* is the most important pathogen of this class.

(c) *Soft rots*. These are caused by bacteria which secrete pectinases and other tissue-macerating enzymes, e.g. *Erwinia carotovora* and *E. chrysanthemi*.

(d) Hypertrophy. Galls and cankers result from proliferation of plant cells induced by certain bacteria, e.g. *Agrobacterium tumefaciens* and *P. syringae* pv *savastanoi*. Plant hormones (auxins and cytokinins) produced by the bacteria are important in stimulating these growths.

(e) *Teratomas*. Some agrobacteria cause production of adventitious shoots or roots. Many mycoplasmas cause floral parts to be replaced by leaf-like structures (phyllody).

(f) *Chlorosis*. This often accompanies other symptoms, e.g. leaf spots may be surrounded by a yellow zone, sometimes caused by toxins.

(g) *Hypersensitivity* (not strictly a disease symptom). This is a resistance response to invading avirulent pathogens and consists of rapid, localized necrosis of plant tissues, creating a hostile environment for the bacteria (15). While hypersensitivity has been extensively studied, it must be stressed that it is not the only manifestation of incompatible plant–pathogen interactions (16).

Our knowledge of the biochemical determinants of disease is very limited. Extracellular enzymes capable of degrading plant tissues, phytotoxins, plant growth regulating substances, and extracellular polysaccharides are produced by many plant pathogenic bacteria, but genetic studies have shown that many other factors are essential for pathogenicity (3). The goal of drawing up a complete catalogue of virulence determinants is still a long way from being achieved.

Unlike fungi, bacteria have no active penetration mechanisms for gaining entry into plants, and are dependent on the presence of natural or induced holes, such as stomata, hydathodes, wounds caused by abrasion or by insects, etc. The tissues colonized depend to a considerable extent on the portal of entry. Bacteria which enter stomata colonize intercellular leaf spaces, whereas those, such as *X. campestris* pv *campestris*, which enter hydathodes gain access to the vascular system, while mycoplasmas and spiroplasmas inhabit only phloem following deposition by phloem-feeding leafhopper vectors. Plant pathologists are most interested in the behaviour of bacteria inside plants, but the ability of the organisms to colonize and survive in other habitats is of great importance in epidemiology. Apart from mycoplasmas, spiroplasmas, and some xylem-limited fastidious bacteria, specific relations with insect vectors are rarely found. Some pathogens such as *Agrobacterium* spp. and *P. solanacearum* may survive for long periods in soil, whereas *Xanthomonas* spp. can only survive in soil in association with plant material. Many pseudomonads and xanthomonads are well adapted to growth and survival on plant surfaces and the epiphytic populations serve as inocula to infect the plant when conditions become suitable. Initial colonization of a plant can result from:

- growth in contaminated soil;
- contaminated seed;
- transfer of bacteria in aerosols or raindrops driven by wind from other infected plants, including weed hosts;
- insect vectors in the case of mycoplasmas.

3. Culturing plant pathogenic bacteria

Gram-negative plant pathogenic bacteria are easily cultured aerobically in both complex and defined laboratory media, usually in the temperature range 20–32°C. Many media formulations and techniques have been published, and as an illustration the procedures employed in the author's laboratory for *X. c. campestris* will be outlined. The basic liquid medium is NYGB (17), containing peptone (5 g/litre), yeast extract (3 g/litre) and glycerol (20 g/litre). Add agar to 1.0 g/litre for NYGA plates. Use of glycerol as a carbon source gives less extracellular polysaccharide and therefore more compact colonies, conversely in studies of polysaccharide production NYG medium can be supplemented with 20 g/litre glucose. Grow cultures for daily use on stock plates of NYGA. After incubation for two days, seal with plastic tape and store in a refrigerator. Inoculate liquid cultures with a loopful of bacteria from the stock plate and grow with shaking in a New Brunswick orbital incubator at 100 r.p.m. at 28–32°C.

The stock plates may be used for several weeks before viability is lost. Although fresh stock plates can be prepared by subculture from the previous plates, it is unwise to repeat the process more than two to three times because of the risk of selecting variants of altered pathogenicity. It is better to return to primary stored cultures. These may be either:

- lyophilized cells stored *in vacuo*, which is excellent but inconvenient and time-consuming, or
- liquid cultures in NYGB stored frozen at $-20°C$ or $-70°C$ with a cryoprotectant such as glycerol (20%) or dimethylsulphoxide (10%).

Chemically-defined media are required for some experiments in genetics and physiology, for example when induction of enzymes by specific inducers is being studied. We use MMX medium, derived from a standard *E. coli* formulation (17). *X. c. campestris* grows slowly in this medium, but more rapidly if supplemented with amino acids.

Gram-positive bacteria are generally more slow-growing. Spiroplasmas need special techniques and media (7) and non-helical mycoplasmas cannot yet be cultured *in vitro*.

4. Testing pathogenicity

The choice of plant inoculation procedure for testing the pathogenicity of a bacterial strain is often a compromise between the desirable and the feasible. Ideally the route of introduction into plants should be the same as that which occurs predominantly in nature, but in many cases this may involve procedures which are too time-consuming for routine use. Genetic studies, for example the isolation of mutants (18–21) or the search for cloned DNA which complements a mutant (22), may require the testing of thousands of individual bacterial colonies for pathogenicity, so a simple, rapid testing procedure is essential. In devising such simplified methods a degree of 'naturalness' may have to be sacrificed, resulting either in short-circuiting of part of the normal infection cycle, or even the creation of a new disease as a result of introduction of bacteria into tissues which they do not normally colonize. Our experience with *X. c. campestris* summarized in *Table 2*, provides an instructive example by comparison of four inoculation methods.

Table 2. Comparison of methods for inoculating *X. c. campestris* into plants

Characteristic	Method[a]			
	1	**2**	**3**	**4**
Simple and rapid	+ +	−	−	+ +
Reliable result	+ +	+ +	+	+ +
Suitable for mass screening of bacterial clones	+ +	−	−	+ +
Applicable to many varieties/species	+ +	+ +	+ +	−
Realistic model of natural disease	−	−	+ +	+ +
Apparent pathogenicity of protease- or polysaccharide-deficient mutants	+	+	−	−

[a] 1 = Stabbing aseptically-grown seedlings. 2 = Syringe infiltration of suspensions into intercellular spaces of leaves. 3 = Application of suspensions to incisions at leaf margins. 4 = Soaking seeds in bacterial suspensions before planting. See text for full description of inoculation methods.

(a) 5–7-day old aseptically grown turnip seedlings are inoculated by touching a fine sharp needle on a bacterial colony and then stabbing through the hypocotyl (18). Symptoms of browning and rotting develop as the bacteria multiply over a 3–4 day period.

(b) A plastic syringe (without a needle) containing a bacterial suspension is pressed gently but firmly against the abaxial surface of a mature turnip leaf, which is supported on the other side by the fingers of the operator. Pressure on the plunger of the syringe forces the suspension through stomata into the intercellular space of the leaf, where the bacteria multiply and give a chlorotic, necrotic lesion after 5–7 days (23).

(c) The top of a mature turnip leaf is submerged in a bacterial suspension and small nicks are made at the edges of the leaf with a sharp scalpel (24). After about one week wedge-shaped chlorotic and necrotic lesions are visible which gradually enlarge and progress into the leaf panel. These symptoms are similar to the natural black-rot disease.

(d) Surface-sterilized radish seeds are soaked in a bacterial suspension and then planted out. The bacteria colonize the surface of the developing seedling and subsequently enter cotyledons and leaves, probably through stomata. Symptoms of chlorosis and necrosis develop after 7–10 days (24). A miniaturized version of this method has been developed for mass screening purposes, e.g. isolating mutants (manuscript in preparation).

The advantages and disadvantages of these methods are indicated in *Table 2*. Methods (a) and (b) are unnatural because bacteria are introduced into tissues not normally colonized in field -infected plants. Method (c) is more satisfactory as only the epiphytic colonization and hydathode penetration stages of the natural infection process (25) are bypassed, but the procedure is too cumbersome for mass screening. Method (d) is an excellent model of natural seedling infection, and with appropriate modifications the onset of symptom development could be delayed until fully expanded leaves are present.

Conclusions drawn from experiments using simplified pathogenicity tests must be treated with caution. *X. c. campestris* mutants defective in synthesis of protease or extracellular polysaccharide appear to retain pathogenicity in seedling-stab or leaf-infiltration tests ((a) and (b)), but in the more natural tests ((c) and (d)) they show much reduced virulence or aggressiveness (unpublished observations).

5. Genetic methodology

The suitability of Gram-negative bacterial pathogens for genetic investigations is one of the main reasons for the upsurge in research interest in these organisms in recent years, and techniques have been devised for applying all the major elements of modern genetic methodology to these organisms. Reference 26 gives a comprehensive review of the topic.

5.1 Mutagenesis

Mutagenesis can be induced with chemical and physical agents or with transposons (27). Chemical mutagenesis or irradiation protocols established primarily for *E. coli* (28) can be optimized to suit the particular organism. Transposon mutagenesis is often preferred because mutations usually give complete loss of function (and therefore clearly-recognizable phenotypes) and because subsequent cloning of the gene is facilitated by using the transposon as a selectable genetic and physical marker for the disrupted gene (i.e. the mutant allele). Transposon mutagenesis demands a suitable delivery vehicle such as a 'suicide' plasmid, and several general systems have been developed from Gram-negative bacteria (27, 29). Experience with many plant pathogens indicates that selection of the best system is empirical; considerable inter-strain variation is found, and some pathogens may be refractory to generalized mutagenesis (17). A useful form of transposon mutagenesis for analysing localized regions of the genome uses 'marker-exchange' (30, 31) to transfer a mutation from pre-mapped cloned DNA into the corresponding site in the genome.

The simple growth requirements of the Gram-negative pathogens allow recovery of a wide variety of auxotrophs, as well as the isolation of mutants altered in pathogenicity.

5.2 Genetic mapping

Mutations can be mapped by conjugation using certain plasmids as sex factors (26). However, mapping has not been fully exploited, probably because it has been overshadowed by recombinant DNA technology which became generally available shortly after the development of 'classical' techniques for pathogens.

5.3 Gene transfer

The three methods of gene transfer in bacteria, namely conjugation, transduction, and transformation, have all been applied to pathogens (26). Chapter 5 gives details of transformation methods. Conjugation is particularly important for the transfer of cloned DNA carried in broad host range cloning vectors into pathogens. Typical protocols are given in refs 17 and 22.

5.4 Gene cloning

Procedures for cloning and manipulating genes of plant pathogens do not differ substantially from those now used so widely for a large number of organisms (32). The most important aspect is the availability of broad host range cloning vectors which can be transferred to, and stably maintained in, almost any Gram-negative bacterium. Thus, genes cloned and 'stored' in a convenient host such as *E. coli* can be moved into strains of the pathogen for functional tests in the correct genetic environment. The vectors are derived from broad host range plasmids of incompatibility groups P, Q, and W (33–35), and many general and special-purpose variants have been constructed.

References

1. Smith, E. F. (1905, 1911, 1914). *Bacteria in relation to plant disease*, Vols 1–3. Carnegie Institute, Washington.
2. Bradbury, J. F. (1986). *Guide to plant pathogenic bacteria*. CAB International, Farnham Royal.
3. Daniels, M. J., Dow, J. M., and Osbourn, A. E. (1988). *Annu. Rev. Phytopathol.*, **26**, 285.
4. Melchers, L. S. and Hooykaas, P. J. J. (1987). In *Oxford surveys of plant molecular and cell biology*, Vol. 4 (ed. B. Miflin), p. 167. Oxford University Press, Oxford.
5. Ream, W. (1989). *Annu. Rev. Phytopathol.*, **27**, 583.
6. Whitcomb, R. F. and Black, L. M. (1982). In *Plant and insect mycoplasma techniques* (ed. M. J. Daniels and P. G. Markham), p. 40. Croom Helm, London.
7. Lee, I. M. and Davis, R. E. (1983). *Appl. Environ. Microbiol.*, **46**, 1247.
8. Kirkpatrick, B. C., Stenger, D. C., Morris, T. J., and Purcell, A. H. (1987). *Science*, **238**, 197.
9. Leyns, F., de Cleene, M., Swings, J. G., and de Ley, J. (1984). *Botan. Rev.*, **50**, 308.
10. Lazo, G. R., Roffey, R., and Gabriel, D. W. (1987). *Int. J. Syst. Bacteriol.*, **37**, 214.
11. Sawczyc, M. K., Barber, C. E., and Daniels, M. J. (1989). *Molec. Plant–Microbe Interactions*, **2**, 249.
12. Whalen, M., Stall, R. E., and Staskawicz, B. J. (1988). *Proc. Natl. Acad. Sci. USA.*, **85**, 6743.
13. Ellingboe, A. H. (1984). In *Advances in plant pathology*, Vol. 2, (ed. D. S. Ingram and P. H. Williams), p. 131. Academic Press, New York.
14. Keen, N. T. and Staskawicz, B. J. (1988). *Annu. Rev. Microbiol.*, **42**, 421.
15. Klement, Z. (1982). In *Phytopathogenic prokaryotes*, Vol. 2 (ed. M. S. Mount and G. H. Lacy), p. 149. Academic Press, New York.
16. Collinge, D. B. and Slusarenko, A. J. (1987). *Plant Molec. Biol.*, **9**, 389.
17. Turner, P., Barber, C., and Daniels, M. (1984). *Molec. Gen. Genet.*, **195**, 101.
18. Daniels, M. J., Barber, C. E., Turner, P. C., Cleary, W. G., and Sawczyc, M. K. (1984). *J. Gen. Microbiol.*, **130**, 2447.
19. Anderson, D. and Mills, D. (1985). *Phytopathology*, **75**, 104.
20. Boucher, C., Barbaris, P., Trigalet, A., and Demery, D. (1985). *J. Gen. Microbiol.*, **131**, 2449.
21. Malik, A. N., Vivian, A., and Taylor, J. D. (1987). *J. Gen. Microbiol.*, **133**, 2393.
22. Daniels, M. J., Barber, C. E., Turner, P. C., Sawczyc, M. K., Byrde, R. J. W., and Fielding, A. H. (1984). *EMBO J.*, **3**, 3323.
23. Collinge, D. B., Milligan, D. E., Dow, J. M., Scofield, G., and Daniels, M. J. (1987). *Plant Molec. Biol.*, **8**, 405.
24. Gough, C. L., Dow, J. M., Barber, C. E., and Daniels, M. J. (1988). *Molec. Plant–Microbe Interactions*, **1**, 275.
25. Williams, P. H. (1980). *Plant Disease*, **64**, 736.
26. Chatterjee, A. K. and Vidaver, A. K. (1986). In *Advances in plant pathology*, Vol. 4 (ed. D. S. Ingram and P. H. Williams), p. 1. Academic Press, London.

27. Berg, C. M., Berg, D. E., and Groisman, E. A. (1988). In *Mobile DNA* (ed. D. E. Berg and M. M. Howe), p. 879. American Society for Microbiology, Washington.
28. Miller, J. H. (1972). *Experiments in molecular genetics*. Cold Spring Harbor Laboratory Press, New York.
29. Mills, D. (1985). *Annu. Rev. Phytopathol.*, **23**, 247.
30. Ruvkun, G. B. and Ausubel, F. M. (1981). *Nature*, **289**, 85.
31. Turner, P., Barber, C., and Daniels, M. (1985). *Molec. Gen. Genet.*, **199**, 338.
32. Sambrook, J., Fritsch, E. F., and Maniatis, T. (1989). *Molecular cloning: a laboratory manual* (2nd edn.). Cold Spring Harbor Laboratory Press, New York.
33. Franklin, F. C. H. (1985). In *DNA cloning: a practical approach*, Vol. 2 (ed. D. M. Glover), p. 165. IRL, Oxford.
34. Keen, N. T., Tamaki, S., Kobayashi, D., and Trollinger, D. (1988). *Gene*, **70**, 191.
35. Liu, Y. N., Tang, J. L., Clarke, B. R., Dow, J. M., and Daniels, M. J. (1990). *Molec. Gen. Genet.*, **220**, 433.

5

Introduction of cloned DNA into plant pathogenic bacteria

NOEL T. KEEN, HAO SHEN, and DONALD A. COOKSEY

1. Introduction

The introduction of cloned DNA into an organism of interest is an essential part of modern biological experimentation. This may be a vexing problem with poorly characterized organisms containing various uncharacterized plasmids, restriction/modification mechanisms, and recombinational systems. Not surprisingly then, bacterial plant pathogens have often posed difficulties in the introduction of exogenous plasmid DNA. While classical chemical transformation methods have been reported for some of these organisms (see, for example, refs 1 and 2), several pathogenic bacteria have not been successfully transformed or, if so, the frequencies are exceedingly low. Transducing phages are not generally useful with plant pathogens for the introduction of DNA. One clever approach, however, has been the introduction of the cloned *E. coli lam*B gene into other Gram-negative bacteria (3). These cells expressing LamB protein are then receptive to transduction with packaged phage lambda cosmid clones. Since phage packaging is required, this is not a general approach for the introduction of DNA into plant pathogenic bacteria, but may be of use for the establishment of cosmid libraries.

The most popular method to introduce DNA into plant pathogenic bacteria has been conjugational matings. The use of helper plasmid (4) or chromosomal (5) mobilization factors *in trans* has permitted the transfer of broad host range cloning plasmids from *E. coli* into a range of recipient bacteria, including plant pathogens. However, some bacteria conjugate with very poor efficiency or not at all. Conjugational methods also require that recipient bacteria carry antibiotic markers to select against the donor bacteria, and problems may accordingly arise in the isolation of desired transconjugants. Nevertheless, its simplicity makes conjugation an appealing method.

In view of the above problems with chemical transformation and conjugational methods for introducing DNA into certain bacteria, attempts have been made to utilize electroporation techniques. Eukaryotic cells have routinely been transformed by electroporation for several years, but only recently have bacteria been successfully transformed at reasonable frequencies using this method (6, 7). The technology involves mixing competent cells with plasmid DNA and exposing them to a rapid electrical discharge at high field strength which presumably causes transitory pores in cell walls/membranes, permitting DNA uptake. After a grow-out period in medium without antibiotics, the transformed cells are then plated on desired selection media. Electroporation conditions for individual bacteria require optimization of parameters and transformation efficiency reflects the genetic background of the recipient bacteria. With restriction and recombination deficient strains of *E. coli*, transformation efficiencies exceeding 10^{10} colonies per μg of DNA have been achieved (7). This is higher than the usual efficiency with chemical methods, although recent improvements in this area result in efficiencies routinely exceeding 10^7 colonies per μg DNA (8). Gram-positive *Bacillus* spp. have also recently been electroporated with high efficiency (9). Based on these successes, we experimented with electroporation for the transformation of several plant pathogenic bacteria, particularly those that have been difficult or impossible to transform with conventional conjugation or chemical transformation techniques. It is our opinion that electroporation has considerable potential for such organisms. In this chapter, we outline a generalized protocol for electroporation that has been successfully used with several plant pathogenic bacteria and we also present a general outline for the introduction of DNA into these bacteria by conjugation.

2. Plasmids used for conjugation and electroporation

Shuttle plasmids that will replicate in both the organism of interest as well as *E. coli* are useful since constructs may be prepared in *E. coli* and then introduced into the plant pathogen. Many such vectors are available, based on several different incompatibility groups. We have had considerable success with relatively small *inc*P or *inc*Q plasmids containing polylinkers for cloning as well as antibiotic selection markers that are expressed in many Gram-negative bacteria (10). We have also electroporated indigenous *Pseudomonas* plasmids conferring copper resistance (11).

3. Electroporation

We have electroporated several plasmids into members of three genera of Gram-negative bacterial plant pathogens (*Table 1*). Various washing protocols for the development of competent cells and varied electroporation parameters as well as the amount of DNA in the transformation mixtures have been assessed. The relatively simple method presented in *Protocol 1* has yielded significant numbers of colonies with all Gram-negative plant pathogenic bacteria tested (*Table 1*), but greater efficiency could doubtless be achieved through optimization of parameters for each bacterium. However, our objective was to develop a

Table 1. Bacterial strains electroporated

Strain	Plasmid[a]	Selection[b]	Efficiency[c]
Pseudomonas syringae pv syringae PS61	pPSI1	Cu	5×10^4
P. s. pv syringae PS61	pPT23D	Cu	5×10^4
P. s. pv tomato PT23	pDSK519	Km	5×10^4
P. s. pv tomato PT23	pRK415	Tc	7×10^2
P. s. pv glycinea R4	pDSK519	Km	7×10^3
P. cichorii 0284-10	pooled pLAFR3 cosmid clones	Tc	1×10^5
Xanthomonas campestris pv vesicatoria 078518	pooled pLAFR3 cosmid clones	Tc	1×10^2
Erwinia chrysanthemi EC16	pDSK519	Km	1×10^4

[a] Plasmid pPSI1 is a 47 kb plasmid isolated from *P. syringae* which confers copper resistance to recipient bacteria; pPT23D is from *P. syringae* pv *tomato* and confers copper resistance (11); pLAFR3 was constructed by Staskawicz et al. (15); other plasmids are described in Keen et al. (10).
[b] Tc = tetracycline at 25 μg/ml; Km = kanamycin at 50 μg/ml; Cu = copper sulphate at 1 mM.
[c] Colonies recovered per μg DNA transformed.

generalized protocol that could be expected to yield significant numbers of transformants with several different pathogenic bacteria.

Protocol 1. Electroporation method

1. Grow bacteria overnight in shaking culture at 28–30°C (use King's medium B; KMB; ref. 12) for pseudomonads and xanthomonads and LB (13) for erwinias). Transfer the cells to fresh medium and continue growth for 3–6 h to an absorbance at 500 nm of 0.5 to 1.0 (c. 10^9 cells/ml).

2. Cool the culture on ice and pellet the cells by centrifugation in sterile 30 ml Oak Ridge tubes or other suitable tubes for 2 min at 7000 g; decant off the medium and suspend the cells by vortexing vigorously in an equal volume of ice-cold 0.5 M sucrose in distilled water (sterilized by autoclaving). Again pellet the cells by centrifugation and wash twice more with cold 0.5 M sucrose.

3. Resuspend the final pellet in a volume of cold 0.5 M sucrose equivalent to c. half the original culture volume. Cells may be kept on ice for several hours. We have also frozen cells in 0.5 M sucrose for up to 5 weeks at −80°C and, following thawing on ice, used them for electroporation. Frequencies of transformation, however, were only 1–10% those of freshly prepared cells.

4. Add 2 μl of DNA (1–100 ng of total plasmid DNA, preferably purified by caesium chloride density gradient centrifugation) to 100 μl of washed cells in 0.5 M sucrose, mix gently, keep on ice and electroporate within 30 min.

5. We use the Bio-Rad Gene Pulser apparatus with Pulse Controller accessory. Electroporation conditions may vary by organism,[a] but initially use 0.2 cm disposable Bio-Rad cuvettes[b] and settings of 7.5–10 kV/cm, 200 ohms, and

Protocol 1. *Continued*

25 μF. Add mixtures of DNA and bacteria to the cuvettes (which are at room temperature) and immediately electroporate.

6. After electroporation, immediately transfer the bacteria to a microcentrifuge tube, mix with 0.9 ml of the medium (without antibiotics) used to grow the bacteria and incubate at 28°C, with shaking, for 1–3 h.

7. Plate aliquots of the bacterial suspension on to agar medium supplemented with desired antibiotics and incubate at 28°C.

8. Plasmid DNA extracted from plant pathogenic bacteria by rapid methods generally does not produce good results following restriction and electrophoresis. Assuming the transformed plasmid will replicate it *E. coli*, extract plasmid DNA from several random transformants by the rapid boil method (14) and transform *E. coli* cells using chemical or electroporation methods. Extract plasmid DNA from *E. coli* transformants and characterize by restriction digestion to ensure no alteration has occurred in the plant pathogenic bacterium.

[a] The manufacturer circulates useful comments concerning modified parameters for various specific bacteria.
[b] The disposable Bio-Rad cuvettes can be re-used many times by rinsing with 70% ethanol (2 times) followed by sterile distilled water (3 times) between transformations.

---

4. Conjugation

A general method for the introduction of broad host range plasmids into plant pathogenic bacteria is presented in *Protocol 2*. We have successfully used this method with several *Pseudomonas syringae* pathovars that are marked with rifampicin and/or chloramphenicol resistance. Most *P. syringae* pathovars are also naturally resistant to ampicillin, permitting the selection of transconjugants on triple antibiotic medium. Isolating transconjugants from donor cells can be a problem, particularly with recipients marked only with rifampicin, because spontaneous *E. coli* mutants can be selected. The use of minimal media may also aid in selection against the donor *E. coli* strains and diagnostic characteristics of the recipient, such as the fluorescence of certain pseudomonads on KMB medium, help ensure the strains of *P. syringae* or *X. campestris* used. Conjugation frequencies vary widely, from 10^{-3} to 10^{-8} per donor bacterium; the latter value is of marginal utility. Problems may also arise due to incompatibility of the introduced plasmid with indigenous plasmids, but these can be solved by using cloning plasmids from different incompatibility groups.

It has been suggested that higher temperatures during matings may limit restriction enzymes in the recipient bacteria while incubation at 4°C can discriminate against growth of the donor. Finally, construction of *recA* strains of the recipient bacteria may reduce problems with rearrangement of introduced plasmids.

Protocol 2. Conjugation method

1. Grow recipient bacteria on agar plates (generally King's medium B) supplemented with appropriate antibiotics. It is important to use rapidly growing cells, so for bacteria that are difficult to conjugate, grow the cells in broth and harvest at mid-log phase.

2. Grow *E. coli* donor cells (and helper plasmid cells if triparental matings are used) on L agar plates at 37°C with required antibiotics.

3. Resuspend the various cells in sterile water to give an absorbance at 600 nm of *c.* 1.0.

4. Spot 20 μl of recipient cells on to the surface of an agar mating plate (generally King's medium B) and briefly allow to dry in a transfer hood. Overlay the first spot with 20 μl of the *E. coli* mobilization/plasmid donor strain (such as SM10 (5)) carrying the plasmid to be introduced. For triparental matings, add 20 μl of *E. coli* cells carrying the plasmid to be moved and, once this has dried, overlay with 20 μl of a helper strain such as *E. coli* HB101 carrying pRK2013 (4).

5. Incubate the plate overnight at a temperature favouring growth of the recipient.

6. Transfer cells from the mating plate and resuspend in sterile water. Plate several serial dilutions on to plates of antibiotic selective medium to obtain single colony transconjugants.

7. Extract DNA from the recipient bacteria with the rapid boil method (14) and, following isopropanol precipitation and drying of the pellet, redissolve the DNA and transform *E. coli*. Extract DNA and analyse the plasmids by restriction analysis to ensure against alteration of plasmids during maintenance in the recipient organism.

5. Discussion

Conjugation has been the most popular method for introducing plasmid DNA into bacteria that lack high efficiency transformation systems. However, the general success observed with electroporation of a wide range of Gram-negative bacteria, including plant pathogens (16), indicates that this method will become the favoured method. Plasmids which could not be introduced into certain bacteria by conjugation (e.g. pPSI1 into *P. syringae* pv *syringae*, *Table 1*) or at extremely low frequencies (all plasmids tested into *P. syringae* pv *tomato*) were readily transformed by electroporation (*Table 1*). However, tetracycline resistance plasmids such as pRK415 generally transformed at lower frequencies than those encoding resistance to copper sulphate or kanamycin (*Table 1*). While transformation frequencies for individual bacteria can doubtless be increased by alteration of parameters, we found that washing with 0.5 M sucrose was a good general method for preparation of competent bacteria. The addition of divalent ions, polyethylene glycol, glycerol, or dimethyl sulphoxide did not improve transformation frequencies. We have not experimented with Gram-positive bacteria, but recent successes with *Bacillus* spp. (9) suggest that electroporation may also work with Gram-positive plant pathogens.

Finally, electroporation has been used to cure plasmids from *E. coli* (17). We have also observed the selective curing of certain indigenous plasmids from phytopathogenic bacteria using electroporation. A 100 kb copper resistance plasmid was selectively cured from a strain of *Xanthomonas campestris* pv *vesicatoria* that contained three large plasmids (Voloudakis and Cooksey, unpublished). The 35 kb copper-resistance plasmid pPT23D was also selectively cured from strain PT23 of *Pseudomonas syringae* pv *tomato* by electroporation, but three other plasmids remained (Jasalavich and Cooksey, unpublished). Plasmid curing by electroporation therefore constitutes an attractive alternative to the use of mutagens or other drastic conditions that may cause secondary mutations in bacteria.

References

1. Gross, D. C. and Vidaver, A. K. (1981). *Can. J. Microbiol.*, **27**, 759.
2. Murooka, Y., Iwamoto, H., Hamamoto, A., and Yamauchi, T. (1987). *J. Bacteriol.*, **169**, 4406.
3. Ludwig, R. A. (1987). *Proc. Natl. Acad. Sci. USA*, **84**, 3334.
4. Ditta, G., Stanfield, S., Corbin, D., and Helinski, D. R. (1980). *Proc. Natl. Acad. Sci. USA*, **77**, 7347.
5. Simon, R., O'Connell, M., Labes, M., and Puhler, A. (1986). *Meth. Enzymol.*, **118**, 640.
6. Wirth, R., Friesenegger, A., and Fiedler, S. (1989). *Molec. Gen. Genet.*, **216**, 175.
7. Dower, W. J., Miller, J. F., and Ragsdale, C. W. (1988). *Nucleic Acids Res.* **16**, 6127.
8. Chung, C. T., Niemela, S. L., and Miller, R. H. (1989). *Proc. Natl. Acad. Sci. USA*, **86**, 2172.
9. Schurter, W., Geiser, M., and Mathe, D. (1989). *Mol. Gen. Genet.*, **218**, 177.
10. Keen, N. T., Tamaki, S., Kobayashi, D., and Trollinger, D. (1988). *Gene*, **70**, 191.
11. Cooksey, D. A. (1987). *Appl. Environ. Microbiol.*, **53**, 454.
12. King, E. O., Ward, M. K., and Raney, D. E. (1954). *J. Lab. Clin. Med.*, **44**, 301.
13. Maniatis, T., Fritsch, E. F., and Sambrook, J. (1982). *Molecular cloning: a laboratory manual.* Cold Spring Harbor Laboratory Press, New York.
14. Crouse, G. F., Frischauf, A., and Lehrach, H. (1983). *Meth. Enzymol.*, **101**, 78.
15. Staskawicz, B., Dahlbeck, D., Keen, N. T., and Napoli, C. (1987). *J. Bacteriol.*, **169**, 5789.
16. Mersereau, M., Pazour, G. J., and Das, A. (1990). *Gene*, **90**, 149.
17. Heery, D. M., Powell, R., Gannon, F., and Dunican, L. K. (1989). *Nucleic Acids Res.*, **17**, 10131.

6

RFLP analyses and gene tagging for bacterial identification and taxonomy

DEAN W. GABRIEL and ROBERT DE FEYTER

1. Introduction

Accurate identification of bacterial isolates to species, subspecies, and pathovar groups coupled with an objective bacterial taxonomy are of paramount importance to microbiologists and plant pathologists. Highly refined genetic technologies facilitate high resolution distinctions such that taxonomies based on phenotype (phenetic taxonomies) are proving to be artificial by genetic standards. The availability of powerful analytical tools such as DNA sequencing, DNA–DNA hybridizations, multilocus isozyme analyses, and restriction fragment length polymorphism (RFLP) hybridization analyses is influencing change towards phylogenetic based taxonomies. Accurate strain identification is also important for epidemiological and ecological monitoring purposes. The simplest and most direct method for the short-term monitoring of known strains released into the field is by direct examination of RFLPs of plasmid DNAs. Plasmid instability may, however, prove a problem to long-term monitoring and other methods, such as monoclonal antibody batteries, polymerase chain reaction analyses, specific DNA probe analyses, or the use of genetically marked strains may prove to be superior. When large numbers of samples must be processed the latter two methods may be the best choices. The selection of antibodies, primers, probes, and/or marking strains requires significant preliminary work. This chapter discusses the use and limitations of RFLP technology and genetically marked strains for taxonomy, identification, and epidemiology of plant pathogenic microbes.

2. Phylogeny, taxonomy, and strain identification of unknown isolates

The identification of unknown isolates for diagnostic purposes requires phenotypes that are stable, reliable, and based on taxonomy which is biologically meaningful. Indeed, when molecular techniques that examine phylogenetic data are used for identification, taxonomies based on a standard microbial phenotypic test may be inconsistent and appear arbitrary. Meaningful biological distinctions may also be lost when genetically diverse microbes are classified into a single species 'group' because simple differential phenotypic tests are unavailable. For example, a phenotype-based taxonomic revision of the genus *Xanthomonas* was proposed in 1972 in which over 100 different pathogenic species were condensed into five species, with *X. campestris* containing most strains of the old species (1). Despite the strong evidence that this proposal was artificial (the phenetic data were known to be unrepresentative of the genetic diversity (2), this taxonomy became widely accepted (3). Similarly artificial taxonomic groupings (relative to genetic diversity) have been observed with *Erwinia* strains (4). To allay the concern expressed by many, the term 'pathovar' (short for *patho*genic *var*iant) was proposed for use in indicating pathogenic information (5).

The pathovar designation has no standing in bacterial taxonomy; strains are classified by pathovar, primarily on the basis of a host phenotype (disease symptoms) on the host from which they were first isolated (6), and not on the basis of their own intrinsic properties as bacteria. Some *X. campestris* strains have more than one host and thus pathovar identification can be incidental and arbitrary; for example, *X. c.* pv *alfalfae* attacks both beans and alfalfa plants, and could therefore be named after either host. (Similar observations have been made regarding the nomenclature of some *Rhizobium* strains (7)). One destructive consequence of the artificial taxonomy was that three very different microbes were grouped into *X. campestris* pv *citri*: *X. citri* (ex Hasse), causing the damaging disease '*Asiatic citrus canker*'; *X. c.* pv *aurantifolii*, causing the disease 'False canker'; and *X. c.* pv *citrumelo*, causing 'citrus bacterial spot' (8). There is even evidence that some non-pathogenic xanthomonads, isolated concurrently with a fungal disease of citrus, were misidentified as *X. c.* pv *citri* (9)! The lack of meaningful or appropriate standards for speciation of *Xanthomonas* has been costly; 'citrus bacterial spot' was misdiagnosed as 'citrus canker', a move that cost well over $25 million in eradication efforts (10) and may cost an additional $28 million to settle lawsuits resulting from the misdiagnosis (11). A phylogenetically valid taxonomy could have prevented these costly errors.

It is now widely accepted that:

• the complete bacterial DNA sequence should be the reference standard to determine phylogeny;

• phylogenetic relationships should determine taxonomy;

• nomenclature should agree with (and reflect) genomic information (12).

Partial DNA sequence analyses of 16S rRNA, which evolves very slowly, is useful for the clarification of phylogenetic relationships from the kingdom to the genus level (13) but 16S rRNA may be too highly conserved to be generally useful for discrimination at the species level. Multilocus isozyme analyses (14), DNA–DNA reassociation (12), DNA–RNA hybridization (15), and RFLP analyses (8) approach the sequence standard, and reveal variation at a level that has been used to distinguish bacterial species. These techniques are not generally useful at taxonomic levels above the species, since sequence diversity is generally too high for detectable hybridization. *Figure 1* shows a comparison of the useful range of such molecular genetic methods, based upon published use of these techniques. A widely accepted phylogenetic standard is that strains which have 70% or greater relatedness by DNA–DNA reassociation are genetically the same species (12). The 70% criterion was determined empirically by analysing DNA hybridization data for clusters, and finding DNA reassociation values of essentially all bacteria studied forming natural clusters in the 80–100% range (16). Strains with DNA hybridization values below 70% also form heteroduplexes with substantially reduced melting temperatures (T_m), implying many differences in the overall DNA sequences (17). Any chosen value may be somewhat arbitrary, but at least classifications based on a phylogenetic standard will be more consistently predictive of the biological properties of the organisms.

There are some notable exceptions to the 70% standard for speciation based on DNA–DNA reassociation. For example, *Escherichia coli* and the four species of *Shigella* cannot be separated on the basis of DNA–DNA reassociation (18); *Bordetella pertussis* apparently differs from *B. parapertussis* by mutations in one or a very few genes (19). The significant pathological differences in these strains make separate species names highly desirable as would the separation of certain economically devastating plant pathogens into separate species. In these exceptional cases, nomenclature would not necessarily follow phylogeny. Identification of such strains to species may require a diagnostic test for the specific pathogenicity factor that induces the disease phenotype. For the vast majority of microbes, however, identification may be based on any phenotype that stably reflects the phylogenetic differences upon which the taxonomy is based.

Detailed protocols for determining DNA–DNA reassociation and DNA–RNA hybridization values (20) and multilocus isozyme analyses (21) have been published elsewhere but it should be noted that different methods for

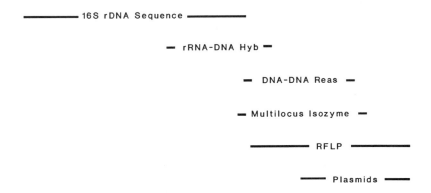

Division Class Order Family Genus Species/Pathovar Strain

Figure 1. Representation of the useful range of phylogenetic techniques, after (37).

determining the DNA–DNA reassociations give different results, particularly at lower reassociation values (15). By the 70% reassociation standard, a large number of strains classified as *X. campestris* may be separated into different species. For example, strains of *X. campestris* pvs *translucens, oryzae, pruni,* and *hyacinthi* have DNA–DNA reassociation rates ranging from 4–15% (22). Similarly, strains of *X. c.* pvs *hyacinthi, pelargonii, carotae, begoniae, vesicatoria, vasculorum, tamarindi, juglandis, manihotis, eranii, oryzae, malvacearum,* and *pruni* exhibit DNA–DNA homology to *X. c.* pv *translucens* in the range of 7–29% (2). However, these pathovars have not, as yet, been reinstated to species, even though other 'recognized' species of *Xanthomonas* show reassociation values in the same range (2). This may be due to the lack of familiarity with these tests because they are impractical for use in routine identifications. Typically, 50–100 μg of carefully prepared DNA is needed for each reference sample and unknown strain, and all possible pair-wise combinations need to be run. As discussed below, RFLP analyses are useful for both taxonomic and identification purposes.

3. RFLP analyses

RFLP analysis provides a routine tool for species and strain identification by giving a very fine level of taxonomic discrimination. RFLPs are widely applicable to phylogenetic and taxonomic questions, having proven useful for cultivated plants (23), wild plants (24), slime moulds (25), yeast (26), plant pathogenic fungi (27), nematodes (28), insects (29), and bacteria (8, 30–35). Results of RFLP analyses of bacterial DNA compare favourably with those obtained using serology (34) DNA–DNA hybridizations and isozyme polymorphism analyses. For example, *Pseudomonas syringae* pv *syringae* strains that shared no common bands with *P. syringae* pv *tomato* strains by RFLP analyses showed only a 37–47% rate of DNA–DNA reassociation (30). Similarly, *X. citri* and *X. phaseoli* have been reinstated to species on the basis of RFLP analyses alone (8) and show 29% and 44% respective rates of DNA–DNA reassociation (spectrophotometric method) to the *X. campestris* type strain (ATCC 33913) (Gabriel, unpublished), and are similarly distant from *X. c.* pv *translucens, X. c.* pv *oryzae,* and *X. c. pruni* (2).

The *quantitative* values of genetic distances obtained in RFLP analyses are dependent upon the probes used, although generally the same *qualitative* relationships are revealed no matter which probes are chosen. Differences in quantitative resolution depend upon the degree of conservation of the loci examined. Some regions of *Xanthomonas* and *Pseudomonas* appear to be much more highly conserved than others (30–32). As with total DNA–DNA or DNA–RNA reassociations, RFLP analyses of identical strains by different investigators, using different DNA probes, usually yields identical relational conclusions, but at different quantitative levels of discrimination (for example, compare ref. 33 with 34). If probes with low discrimination potential are used, lower dissimilarity values will be obtained than if probes with high discrimination potential are chosen. Resolution levels depend entirely on which loci are examined. We have found that chromosomally-derived probes that hybridize to plasmid DNAs are quite suitable for both identification and taxonomic purposes. Probes may include highly repeated DNA sequences (27, 35). Repeated DNA sequences offer the advantage of examining multiple loci, but may also reveal exceptionally high, and perhaps misleading, levels of polymorphisms among certain groups of related strains. The 70% criterion set as a break-point for speciation by DNA–DNA hybridization works in practice because there is a natural separation of most strains of bacteria into clonal groups that occurs below that level (16). For RFLP-based taxonomy, a criterion level may be chosen by selecting genetic loci which reveal meaningful phylogenetic clusters. As far as DNA–DNA reassociations are concerned, the criterion level chosen is arbitrary, but at some level the natural clusters are revealed. For taxonomic as well as strain identification purposes, DNA probes should be used which are capable of high levels of discrimination. For speciation by RFLP analysis, the criterion we suggest for inclusion of strains in a species is at least 80% similarity with the type strain as determined by probes capable of revealing 20% or less similarity between species of a genus.

Analysis by RFLP involves cell lysis, DNA extraction (*Protocol 1*) restriction digestion, size separation of fragments by agarose gel electrophoresis, Southern blotting, hybridization of blotted DNA with probe (*Protocol 2*), visualization of hybridizing bands, production of a similarity matrix, and finally production of a phenogram. While DNA bands may be visualized directly in gels by staining with ethidium bromide, the very large numbers of bands makes subsequent data quantification, even by laser densitometry, practically impossible. Normally only the smallest fragments are analysed, together with any high copy number plasmid DNA bands superimposed on the chromosal fragments. Furthermore, the mere co-migration of two similarly sized fragments does not mean that the fragments are the same. By examining a subset of the chromosome by hybridization in Southern blots, one can readily quantify 'matches', and mismatches visually verify that the fragments matched are homologous. The process is shown in *Figure 2*. The preparation of DNA from phytopathogenic microbes differs somewhat from methods well-established for *E. coli* (for example, see ref. 36).

Protocol 1. Extraction of DNA from xanthomonads

1. Inoculate an 8 ml (80 ml) TY-MOPS (8) culture in a flask of at least 25 ml (250 ml) volume.[a]

2. Grow at 30°C with 'slow' shaking to give cells in mid to late exponential phase.

3. Pellet cells at 5000 g, 10 min in microcentrifuge tubes (40 ml teflon or polypropylene tubes).

4. Wash cells in 1 ml (40 ml) NE buffer (50 mM EDTA, 0.15 M NaCl, pH 8.0),

Protocol 1. *Continued*

   repellet as in step 3. Discard the loose, colourless material 'floating' near the pellet.[b]

5. Repeat step 4. The cells may be frozen at this step.

6. Resuspend the pellet from step 5 in 600 μl (20 ml) of 150 μg/ml proteinase K solution in NE buffer. Add 30 μl (1 ml) 20% SDS. Thoroughly mix by vortexing.

7. Incubate at 50°C, 1 h.

8. Extract with equal volume chloroform/phenol/isoamyl alcohol (24:25:1 v/v/v), buffered with 10 mM Tris–HCl, pH 8.0. Mix by vortexing.

9. Centrifuge at 5000 g for 5 min to separate the layers.

10. Transfer the upper layer to a fresh tube, avoiding the white interface. Repeat step 6 several times until the interface is clear.

11. Add 0.1 vol. 3 M NaAc and one vol. isopropanol. Mix by vortexing.

12. Spool out precipitate with a freshly flame-sealed Pasteur pipette or capillary tube. Rinse by dipping in 70% ethanol. Allow most of the ethanol to drain. Do not dry.[c]

Steps 13 to 15 can be omitted although the DNA may not digest as readily

13. Transfer to clean tube containing 500 μl (6 ml) 10 mM Tris–HCl pH 8.0, 1 mM EDTA, 100 μg/ml RNAase A. Allow DNA to dissolve overnight at 4°C.

14. Add an equal vol. chloroform/isoamyl alcohol (24:1 (v/v)). Thoroughly mix by vortexing. Centrifuge 5 min at 5000 g.

15. Transfer upper phase to a clean tube, and add 0.1 vol. 3 M NaAc, 2 vol. 95% ethanol at room temperature, mix.

16. Spool out DNA, rinse, and drain as in step 12. Transfer to a fresh tube containing 200 μl (2 ml T.E.). Dissolve at 4°C overnight.

---

[a] It is important not to add any sugars or glycerol to the growth medium as the extra carbon source appears to stimulate gum production.

[b] *Xanomonad* cells are yellow but the gum is clear. Polysaccharide gums coprecipitate with the DNA and interfere with restriction digestion. You should see a highly viscous, clear layer separate from the cell pellet; and after removal of the gum by washing, the pellet should be yellow, tight, and dry looking.

[c] Steps 12 and 16: Do not dry DNA completely as it will not resuspend easily. Since the resuspension volumes are relatively large, the trace of ethanol left over by incomplete drying is of little consequence.

   Nucleic acid yields following *Protocol 1* are typically in the range of 0.4–1 mg/ml. DNA minipreps prepared as described will typically have 50–80% RNA contamination (as determined by the hyperchromic shift (20)) despite the RNAase treatment in step 13 and spooling the DNA in step 16. Generally,

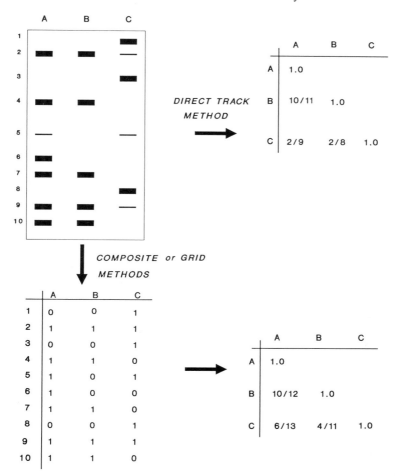

Figure 2. Comparison of two methods of quantifying RFLP data. For simplicity, only a small number of 'major' fragments are illustrated as heavy bands, and minor fragments as light bands. In the Direct Track method, only the most major fragments (up to an arbitrary limit) are used as the positional basis for comparison (denominator), and minor fragments are counted (in the numerator) if they match a major fragment. In the Composit method, major and minor bands are all assigned to the position closest to that of the grid used.

however, all preparations made in one experiment should have similar levels of RNA contamination and therefore a uniform amount of DNA (albeit a smaller amount than indicated) will be loaded per lane. The RNA does not appear to interfere with RFLP analyses.

Protocol 2. Restriction analysis

1. Measure the absorbance of the nucleic acid solution using quartz cuvettes

Protocol 2. *Continued*

with a 1 cm light path in a spectrophotometer at 260 nm. Calculate the nucleic acid concentration according to the formula:

$$\text{DNA concentration (mg/ml)} = \frac{\text{Absorbance at 260 nm} \times \text{dilution factor}}{20}.$$

2. Digest 10 μg in a total vol. of 40 μl using 16 to 24 units of enzyme for 1 h at 37°C. Use an excess to enzyme to ensure complete digestion.

3. Prepare a horizontal agarose gel essentially according to ref. 36. Use 200 ml 0.7% agarose (Seakem GTG, FMC Bioproducts) in Tris–acetate, pH 8.0, buffer for a *c.* 12 inch long gel.

   Take care to ensure reproducibility from gel to gel

   • Ensure agarose is completely dissolved.

   • Replace any water lost by evaporation.

   • Pour gel once agarose has cooled to 47°C in a water bath.

   • Load and run gel within 90 min of pouring or immerse gel in buffer or cover with plastic wrap.

4. Electrophorese 20 μl of the restriction digest at 35 V for 16 h. Also run size markers and 2 tracks of 1 ng probe DNA.

5. Stain the gel in a 0.5 μg/ml solution of ethidium bromide, photograph, and evaluate. The probe samples will not be visible.

The other half of the digest is saved in case some tracks are significantly overloaded. In this case the gel is re-run using appropriate dilutions of these samples. Occasionally some samples appear seriously underloaded, undigested, or partially digested. Such samples should be re-isolated and/or re-digested and the gel re-run.

Protocol 3. Alkaline Southern blot

1. Soon after ethidium bromide staining incubate the gel in 5 vol. 0.25 M HCl for 5 to 10 min.

2. Equilibrate in 3 vol. 0.4 M NaOH for 30 min then transfer to nitrocellulose or nylon membrane[a] by capillary blotting (for example, see ref. 36) overnight with 10 vol. alkaline solution in the transfer tray.

3. Fix DNA to membrane, prehybridize, and hybridize according to manufacturer's instructions with a suitable probe. We use a nick-translation kit (Gibco-BRL, Cat. No. 8160SB) and α-[32]P dCTP (New England Nuclear, Cat. No. NE6-0134). Set up the labelling reaction in a total vol. of 10 μl by mixing:

Protocol 3. *Continued*

- 100–200 ng DNA
- 3 μl ^{32}P-dCTP (30 μCi 3000 mCi/mmol)
- 1 μl cold dNTP mix (dATP, dTTP, dGTP in 10 × buffer)
- 1 μl DNA polymerase

Incubate 60 min, 15°C then immediately fractionate on a 3 inch (*c.* 7.5 cm) Sephadex G-50 column (36).

4. Following *c.* 16 h hybridization wash membrane in 4 × SSC for 5 min and autoradiograph without drying.

5. After a suitable exposure, strip the probe from the membrane as described by the manufacturer. For example for GeneScreen Plus[a] incubate in 200 ml 0.4 M NaOH, 42°C, 30 min then 200 ml 0.1 × SSC, 0.1% SDS, 0.2 M Tris–HCl pH 7.5, 42°C, 30 min.

6. Repeat prehybridization and hybridize with fresh probe.

---

[a] We prefer GeneScreen *Plus* Nylon (New England Nuclear, Cat. No. NEF-976) because DNA probes may be readily stripped from this material after autoradiography and rehybridized several times.

---

4. Similarity coefficients

The most biased step in RFLP analysis lies in deciding which bands are 'matched' or 'not matched'. No two gels will separate DNA fragments at exactly the same positions, and therefore quantitative comparisons cannot currently be made between strains run on different gels. Since DNA fragment mobility is a logarithmic function of size, comparison of different gels may be possible with computer-assisted referencing of bands to appropriate 'stretch' or shrink gels. As fragment position data derived from tracks run on the same gel are subject to bias; it is important for readers and reviewers to carefully examine original autoradiographic data for internal consistency, and to check representative strains to see if the 'quantitated' conclusions appears to reflect the qualitative evidence. Much effort must be spent on this part of the analysis to obtain accurate and reproducible quantitative information.

A commonly used method is to assign a unique number to each and every 'position' in which a band appears in any strain run on the entire autoradiograph (30, 37). This may be accomplished by using a clear overlay to create a composite track containing all bands drawn in. The total number of bands on the composite are counted, assigned numbers, and the composite is compared with each DNA sample track to determine whether or not a band is present in that lane that corresponds to each band on the composite. Increasing levels of bias may occur as the total number of positions on the composite increases, since it becomes increasingly difficult (as band density increases) to decide whether a given band occupies a slightly different position on the composite or corresponds to an already defined position. A variation on this method is to overlay the entire autoradiograph with a grid pattern of parallel lines, perpendicular to the gel tracks. Potential band positions are predetermined by the spacing of the parallel lines, for example every 4 mm. Bands in each track are assigned to one of the predetermined line positions. This method relies on nearly perfect gels and blots, since track skewing or 'smiling' would result in bands from identical strains being assigned to different line positions. Compensation for slight gel imperfections may be made by increasing the grid width, but in so doing, the potential for spurious 'matches' greatly increases. The effect of spurious matches is low for highly similar strains, but may reduce the reliability of data for lower ($<50\%$) levels of similarity. By either the composite or grid methods, band positions data can be digitized and recorded in a 'data matrix'. Comparisons between two strains are made from the data in the matrix. Similarity coefficients between two strains are then calculated using the following estimator of DNA fragment homology (F) (38):

$$F = 2n_{xy}/(n_x + n_y),$$

where n_x and n_y are the number of fragments in each strain, and n_{xy} is the number of fragments shared by the compared strains. The problem of band density may be alleviated by using probes which reveal a relatively low number of (<10) of bands per line, and then using a larger number of DNA probes.

An alternative method is to compare the track data directly between strains (see *Figure 2*). The autoradiogram is cut into strips corresponding to each sample lane, and the lanes are directly compared in pair-wise combinations. A laser densitometer or spectrophotometer equipped with an autoradiogram scanner may be used to localize band positions and determine hybridization intensity. It is convenient to overlay the densitometer traces or the autoradiogram strips pair-wise on a light-box, and simply count the number of bands that match. The ratio of matching bands against the total number in both tracks is an estimate of similarity at the locus examined. This procedure helps to reduce the bias inherent in the 'composite' procedure when a relatively large number of bands (more than 20) are present in each track. Since band matches and mismatches are made directly in side-by-side comparisons between two tracks, and not between each track and a composite of all positions derived from each track run on the gel, there is some improvement in the quality (reduction of bias and improved reproducibility) of the data. This procedure also helps compensate for slight distortions within the gel (slightly skewed tracks or 'smiles').

The assignment of a weighted value to matching pairs of strongly hybridizing bands versus matching pairs of weakly hybridizing bands versus 'strong'–'weak' matches poses a problem. In qualitative comparisons of RFLP analyses with the naked eye, it is obvious that such information is useful in pattern recognition. The simplest procedure (used with the 'composite' protocol) is to weight all bands equally, regardless of intensity. Unfortunately, when a large number of bands are compared on a single gel, the loss of strong versus weak band information strongly biases the comparisons in favour of invalidly high similarity values. We have observed reports in which conclusions were drawn from such 'quantitated' data that were obviously not supported by simple inspection of the autoradiograms. To recover part of this (strong versus weak) information, we suggest the use of a densitometer or spectrophotometer capable of determining peak intensity by integration. We use the Gilford Response II (Ciba Corning Diagnostic Corp.) recording spectrophotometer, which automatically determines the 12 largest peaks in each autoradiograph track. These 12 peaks are counted as 'major' bands, and all other bands (over 12) are counted as 'minor' bands. If the absolute position of a major band in one track matches a major band in the same position (± 1 mm on a 150 mm scale) on the other track, only one band is recorded. In contrast with the 'composite' protocol, minor bands that appear to match minor bands are not recorded. Comparative data from each enzyme–probe combination are totalled, and similarity coefficients between two strains are calculated from the raw data directly using the following estimator of DNA fragment homology (F) (33):

$$F = (n_{xy} + n_{yx})/(n_x + n_y),$$

where n_x and n_y are the numbers of major fragments in strains X and Y, respectively, n_{xy} is the number of major fragments in strain X which match any fragments in strain Y, and n_{yx} is the number of major fragments in strain Y which match any fragments in strain X. Software is required to make the comparisons and select the matches automatically.

Raw totals of matches and total bands are added for each enzyme–probe combination and converted to a percentage of similarity (or dissimilarity). The result is a table of similarity (or dissimilarity) coefficients. A number of algorithms have been written for determining clusters from similarity data; most have used the unweighted pair-group method with averaging (UPGMA). Programs that have been used are NTSYS (39) and Clustan (40). The results of cluster analyses are graphically illustrated as dendrograms. One advantage of Clustan is that the program can tolerate 'missing values'. Although data from separate blots may not be readily combined in a single cluster analysis, it is possible to choose representative strains from within nucleus clusters for separate comparisons in order to determine qualitative relationships between highly clustered groups, in effect, to 'cluster the clusters'. Composite dendrograms may then be drawn which can often be useful for illustrating relationships between large numbers of different strains (8). The limitations here are that relationships between heterogeneous (not tightly clustered) groups cannot be accurately determined, except perhaps by iterative relocation.

One of the advantages of using RFLP technology for identification purposes is that a readily accessible and recognizable pattern 'library' is built up over time. Unknown samples may be processed in the absence of any plant data or reference

DNA. Unlike the quantitative comparisons needed for taxonomic work, qualitative pattern similarities can often be matched to known strains by eye, and later checked, if necessary, by comparative gels. If the host is known, a match or mismatch with suspect strains by simply determining the RFLP pattern with one probe and two enzymes is a virtual certainty.

5. Plasmid DNAs

Not all *Xanthomonas* strains carry plasmids but plasmid DNAs of some species and pathovars may be stable enough to be useful for unknown strain identification (41). *Protocol 4* yields consistently reliable plasmid DNA from *Xanthomonas*.

Protocol 4. Small-scale plasmid preparation from *Xanthomonas* species

1. Grow *Xanthomonas* cultures to mid-exponential growth phase in tryptone–yeast medium (42) or nutrient broth (43). Avoid the use of stationary phase cultures or cultures grown in richer media with excess sugars since these increase xanthan gum production which is not desirable.

2. Harvest the cells from 1 ml by centrifugation in a microcentrifuge tube and remove the supernatant.

3. Wash the cells with 1 ml EDTA/saline (50 mM EDTA, 0.15 M NaCl, pH 8.0), to help remove any gum. Repellet the cells, 5000 g, 10 min.

4. Decant the supernatant, leaving a small droplet of EDTA/saline in the tube. Vortex the tube to resuspend the cells.

5. Add 0.3 ml of CAPS[a] lysing buffer (50 mM CAPS, 3% SDS, 10 mM EDTA, pH 12.4), and mix by taking up once or twice into a Pipetman tip. The mixture should be extremely viscous and clear-yellow.

6. After 2–3 min at room temperature, add 0.15 ml of potassium acetate solution (3 M potassium acetate, titrated to pH 4.8 with glacial acetic acid). Invert the tube briskly several times to mix, cool on ice for 10 min.

7. Centrifuge the tube at *c*. 13 000 g for 3 min. Transfer the supernatant to a fresh tube.

8. Extract the supernatant once with an equal volume of phenol:chloroform: isoamyl alcohol (25:24:1), and once with chloroform:isoamyl alcohol (24:1).

9. Recover nucleic acid from the aqueous phase by precipitation with ethanol. Add 3 vol. of 95% ethanol, invert the tube to mix, and stand at room temperature for 10 min. Pellet the nucleic acid by centrifugation at *c*.13 000 g for 3 min.

10. Decant the supernatant, drain the tube briefly on a tissue, and rinse the tube once with 70% ethanol. Drain again and dry the pellet under vacuum.

Protocol 4. *Continued*

11. Suspend the nucleic acid in 10 μl TE buffer (10 mM Tris–HCl, 1 mM EDTA, pH 8.0). Add 2 μl of 1 mg/ml RNAase (heated to remove DNAase) in water and incubate at 37°C for 30 min.

12. Repeat steps 8, 9, and 10. The second ethanol precipitation requires the addition of 10 μl 3 M sodium acetate, pH 5.2. Take care not to dislodge the DNA pellet when decanting the supernatant.

13. Resuspend the DNA in 20 μl of TE buffer and use 10 μl for analysis.

---

[a] CAPS (3-[cyclohexylamino]-1-propanesulphonic acid), SDS and EDTA are dissolved in water and adjusted to pH 12.0 with NaOH. Do not stir while reading the pH. Allow to stand for 2 h and readjust pH to 12.4. Bring buffer up to volume.

---

6. Gene tagging for known strain identification

For field-release experiments of known strains, it can be difficult to rapidly and quantitatively distinguish between strains released for study and similar naturally-occurring strains (44, 45). One method that has been used is to introduce one or more specific genetic markers. These may be selectable (antibiotic) markers or screenable (visual test) markers, such as β-galactosidase (46), β-glucuronidase (47), or luciferase (48). Perhaps the easiest way to introduce such markers is via transposon mutagenesis (as for example, luciferase (49)). A potential problem with random insertion of these transposons is the possibility that these transposons will somehow find their way out of the target organism and be transferred into another. Despite the well-described mechanisms for horizontal gene transfer among bacteria (transformation, transduction, conjugation, and cell fusion), little is known about the ecological significance of these potential mechanisms (50, 51). There is also the problem of not knowing exactly where the transposon has inserted, and what the result(s) of the random insertion may be at an uncharacterized location. At least one transposon marking system has been developed to avoid such problems. This system involves the use of Tn7, which does not insert randomly in bacterial DNA, but has a specific and well-characterized target site for insertion (52). In this system, developed for marking strains for field release, a gene essential for transposition has been deleted from Tn7 and was recloned for use in marking strains *in trans*.

Another method is to clone marker genes into an already characterized fragment of DNA and then transfer the entire construction into the target organism. Transconjugants or transformants are then screened for retention of the marker gene and loss of vector markers, indicating a double recombination event known as marker exchange. Marker exchange depends upon the loss of the cloning vector, and therefore the vector used may be:

- unable to replicate in the target organisms (e.g. a suicide vector),

- at least somewhat unstable in the target organism (53),
- stable in the target organism, but destabilized in the target after transfer by using a second, incompatible plasmid (see Section 6.1).

6.1 Marker exchange by using an incompatible plasmid

Plasmids from the incompatibility groups *inc*P and *inc*W have been used in marker exchange methods. Such plasmids contain broad host range replication regions and are often capable of replicating stably in a wide variety of Gram-negative bacteria, therefore cloned DNA including a marker gene is stably inherited in the target organisms. To destabilize the replicon and allow selection of the marker exchange products, a second plasmid from the same incompatibility group is introduced into the target organisms which contain the first plasmid which selection is maintained for the marker gene insert. Usually only a small minority of the colonies recovered (1–30%) have lost the first plasmid markers while retaining the marker gene insert, indicating marker exchange has occurred. It is essential in this method that distinguishable markers are used on each plasmid. We have developed a series of *inc*W based plasmids (54) for this method, using pUFR047 (resistant to gentamycin and ampicillin) as the first plasmid, with an *aph*A gene cassette as the marker gene (neomycin and kanamycin resistance) and pUFR049 as the second, destabilizing plasmid (chloramphenicol and streptomycin resistance). These markers are useful in wide variety of xanthomonads. The plasmids can be introduced readily into bacteria either by conjugation or electroporation.

Acknowledgements

We are indebted to Dr Tim Denny for the idea of *Figure 1* and for many fruitful discussions. We also thank Dr Clarence Kado and Dr Isaac Barash for helpful comments on the manuscript.

References

1. Lelliot, R. A. (1972). *Proc. Third Intern. Conf. Plant Pathogenic Bacteria* (ed. H. P. M. Geesteranus), p. 269. Centre for Agricultural Publication and Documentation, Wageningen.
2. Murata, N. and Starr, M. P. (1973). *Phytopath. Z.*, **77**, 285.
3. Bradbury, J. F. (1984). *Bergey's manual of sytematic bacteriology*, Vol. 1 (ed. N. R. Kreig and J. G. Holt), p. 199. Williams and Wilkins, Baltimore.
4. Gardner, J. M. and Kado, C. I. (1972). *Int. J. Sys. Bacteriol.*, **22**, 201.
5. Dye, D. W., Bradbury, J. F., Goto, M., Hayward, A. C., Lelliott, R. A., and Schroth, M. N. (1980). *Review of Plant Pathology*, **59**, 153.
6. Starr, M. P. (1983). *The prokaryotes: a handbook on habitats, isolation, and identification of bacteria* (ed. M. P. Starr), p. 742. Springer, New York.
7. Keyser, H. H., van Berkum, P., and Weber, D. F. (1982). *Plant Physiol.*, **70**, 1626.

Dean W. Gabriel and Robert de Feyter

8. Gabriel, D. W., Kingsley, M. T., Hunter, J. E., and Gottwald, T. R. (1989). *Int. J. Sys. Bacteriol.* **39**, 14.
9. Stapleton, J. J. and Garza-Lopex, J. G. (1988). *Phytopathology*, **78**, 440.
10. Schoulties, C. L., Civerolo, E. L., Miller, J. W., Stall, R. E., Krass, C. J., Poe, S. R., *et al.* (1987). *Plant Disease*, **71**, 388.
11. Longman, P. (1989). *Florida Trend*, **32**, 40.
12. Wayne, L. G., Brenner, D. J., Colwell, R. R., Grimont, P. A. D., Kandler, O., Krichevsky, M., *et al.* (1987). *Int. J. Syst. Bacteriol.*, **37**, 463.
13. Murray, R. G. E. (1984). *Bergey's manual of systematic bacteriology* (ed. N. R. Krieg and J. G. Holt), p. 31. Williams and Wilkins, Baltimore.
14. Selander, R. K. (1985). *Population genetics and molecular evolution* (ed. T. Ohta and K. Aoki), p. 85. Japan Scientific Societies Press, Tokyo.
15. Grimont, P. A. D. (1988). *Can. J. Microbiol.*, **43**, 541.
16. Johnson, J. L. (1973). *Int. J. Sys. Bacteriol.*, **23**, 308.
17. Johnson, J. L. (1984). *Bergey's manual of systematic bacteriology* (ed. N. R. Krieg and J. G. Holt), p. 8. Williams and Wilkins, Baltimore.
18. Brenner, D. J., Fanning, G. R., Miklos, G. V., and Steigerwalt, A. G. (1973). *Int. J. Sys. Bacteriol.*, **23**, 1.
19. Gross, R. and Rappouli, R. (1988). *Proc. Natl. Acad. Sci. USA*, **85**, 3913.
20. Johnson, J. L. (1985). *Methods in Microbiology*, **18**, 33.
21. Selander, R. K., Caugant, D. A., Ochman, H., Musser, J. M., Gilmour, N. N., and Whittam, T. S. (1986). *Appl. Environ. Microbiol.*, **51**, 873.
22. Swings, J., De Vos, P., Van Den Mooter, M., and De Ley, J. (1983). *Int. J. Sys. Bacteriol.*, **33**, 409.
23. Song, K. M., Osborn, T. C., and Williams, P. H. (1988). *Theor. Appl. Genet.*, **76**, 593.
24. Soltis, D. E., Soltis, P. S., Ranker, T. A., and Ness, B. D. (1989). *Genetics*, **121**, 819.
25. Evans, W. B., Hughes, J. E., and Welker, D. L. (1988). *Genetics*, **119**, 561.
26. Pedersen, M. B. (1988). *Modern methods of plant analysis*, p. 180. Springer, Berlin.
27. Hamer, J. E., Farrell, L., Orback, M. J., Valent, B., and Chumley, F. G. (1989). *Proc. Natl. Acad. Sci. USA*, **86**, 9981.
28. Powers, T. O., Platzer, E. G., and Hyman, B. C. (1986). *J. Nematol.*, **18**, 288.
29. Hall, H. G. (1988). *Florida Entomologist*, **71**, 294.
30. Denny, T. P., Gilmour, N. N., and Selander, R. K. (1988). *J. Gen. Microbiol.*, **134**, 1949.
31. Cook, D., Barlow, E., and Sequeira, L. (1989). *Molec. Plant–Microbe Interactions*, **2**, 112.
32. Lazo, G. R., Roffey, R., and Gabriel, D. W. (1987). *Int. J. Sys. Bacteriol.*, **37**, 214.
33. Gabriel, D. W., Hunter, J., Kingsley, M., Miller, J. and Lazo, G. (1988). *Molec. Plant–Microbe Interactions*, **1**, 59.
34. Gottwald, T. R., Alvarez, A. M., Hartung, J. S., and Benedict, A. A. (1991). *Phytopathology*, **81**, 65.
35. Mogen, B. D., Olson, H. R., Sparks, R. B., Gudmestad, N. C., and Oleson, A. E. (1990). *Phytopathology*, **80**, 90.
36. Maniatis, T., Fritsch, E. F., and Sambrook, J. (1982). *Molecular cloning: a laboratory manual*. Cold Spring Harbor Laboratory, New York.
37. Denny, T. (1988). Hand-out presented at a 'Teach In' at the Annual Meeting of the Phytopathological Society, San Diego.
38. Nei, M. and Li, W. H. (1979). *Proc. Natl. Acad. Sci. USA*, **76**, 5269.

39. Rohlf, F. J. Applied Biostatistics, Inc., 3 Heritage Lane, Setauket, New York.
40. Wishart, D. (1987). *Clustan user manual*. Computing Laboratory, University of St Andrews, Scotland.
41. Lazo, G. R. and Gabriel, D. W. (1987). *Phytopathology*, **77**, 448.
42. Beringer, J. E. (1974). *J. Gen. Microbiol.*, **84**, 188.
43. Miller, J. H. (1972). *Experiments in molecular genetics*. Cold Spring Harbor Laboratory Press, New York.
44. Yuen, G. Y., Alvarez, A. M., Benedict, A. A., and Trotter, K. J. (1987). *Phytopathology*, **77**, 366.
45. Gottwald, T. R., Civerolo, E. L., Garnsey, S. M., Brlansky, R. H., Graham, J. H., and Gabriel, D. W. (1988). *Plant Disease*, **72**, 781.
46. Drahos, D. J., Hemming, B. C., and McPherson, S. (1986). *Bio/Technology*, **4**, 439.
47. Jefferson, R. A. (1988). *Genetic engineering: principles and methods* (ed. J. K. Setlow), p. 247. Plenum, New York.
48. Shaw, J. J. and Kado, C. I. (1986). *Bio/Technology*, **4**, 560.
49. Shaw, J. J., Settles, L. G., and Kado, C. I. (1988). *Molec. Plant–Microbe Interactions*, **1**, 39.
50. Trevors, J. T., Barkay, T., and Bourquin, A. W. (1986). *Can. J. Microbiol.*, **33**, 191.
51. Schofield, P. R., Gibson, A. H., Dudman, W. F., and Watson, J. M. (1987). *Appl. Environ. Microbiol.*, **53**, 2942.
52. Barry, G. F. (1988). *Gene*, **71**, 75.
53. Gutterson, N. I., Layton, T. J., Ziegle, J. S., and Warren, G. J. (1986). *J. Bacteriol.*, **165**, 696.
54. de Feyter, R., Kado, C. I., and Gabriel, D. W. (1990). *Gene*, **88**, 65.

7

Fungi

JOHN M. CLARKSON

1. Introduction

1.1 Mode of nutrition

Fungi are heterotrophic eukaryotic microbes which depend on a saprophytic, mutualistic, or parasitic mode of existence. Plant pathogenic fungi represent the most economically important group of microbial pathogens and can be divided into two broad categories on the basis of their nutritional relationship with the host (1). **Necrotrophs** usually cause extensive tissue damage and derive their nutrition from dead or dying cells. These fungi are able to grow, and often complete their life cycle, in axenic culture on simple laboratory media. Because they are capable of a purely saprophytic existence they are referred to as facultative necrotrophs but it is doubtful that this capacity for saprotrophy is fully expressed in the wild, as many pathogens persist as resting structures in the absence of a susceptible host. **Biotrophs** have a more subtle relationship with their hosts, deriving nutrients from living host cells and often causing only minor tissue damage in early infection. They include the downy and powdery mildews, rusts, and smuts and represent many of the most commercially significant pathogens. Structural adaptation to this mode of nutrition is typified by the formation of haustoria—specialized hyphae which penetrate the host cell wall and invaginate the plasma membrane. Although all these pathogens are ecologically obligate biotrophs, the haploid sporidial stage of smut fungi can be readily cultured on simple laboratory media and limited axenic growth of some rusts is possible (2). Some fungi, such as *Phytophthora infestans* and *Colletotrichum lindemuthianum* exhibit a short biotrophic phase of development prior to obtaining nutrients necrotrophically and are referred to as hemibiotrophs.

1.2 Thallus organization

The thallus or vegetative body of a fungus takes the form of either a multinucleate plasmodium lacking a true cell wall (classified in the division *Myxomycota*) or a unicellular yeast-like or filamentous mycelium (classified in the division *Eumycota*). Although many mycologists and plant pathologists follow the classification proposed by Ainsworth (3) and used here (*Table 1*), in which the

Table 1. Classification of the major plant pathogenic fungi[a]

Division Myxomycota (possess plasmodium or pseudoplasmodium)
  Plasmodiophoromycetes—*Plasmodiophora brassicae* (brassica club root), *Polymyxa graminis,* and *P. betae* (cereal and beet root pathogens).
Division Eumycota (unicellular or filamentous mycelium)
  Mastigomycotina (produce motile zoospores)
    Oomycetes—*Pythium* spp. (seedling damping off). *Phytophthora* spp. (mainly root/stem rots), and *Phytophthora infestans* (potato blight). Downy mildews, e.g. *Bremia lactucae* (lettuce), and *Plasmopara viticola* (vine).
  Ascomycotina (sexual spores = Ascospores)
    Plectomycetes—Powdery mildews, e.g. *Erysiphe* spp. (e.g. *E. graminis* (cereals)), *Podosphaera leucotricha* (apple), *Uncinula necator* (vine). *Pencillium* spp. (fruit storage rots).
    Pyrenomycetes—*Ophiostoma ulmi* (Dutch Elm vascular wilt), *Claviceps purpurea* (grain ergot), *Endothia parasitica* (Chestnut blight), *Glomerella* spp. (anamorph: *Colletotrichum*) (anthracnose), *Giberella* spp. (cereal stalk rots), *Magnaporthe grisea* (anamorph: *Pyricularia oryzae) (rice blast)*, *Leptosphaeria maculans* (brassica stem canker).
    Discomycetes—*Sclerotinia sclerotiorum* (stem and root rot), *Pyrenopeziza brassicae* (Brassica light leaf spot).
    Loculoascomycetes—*Ophiobolus graminis* (cereal take-all), *Pyrenophora teres* (barley net blotch), *Cochliobolus heterostrophus* (anamorph: *Helminthosporium maydis* (maize Southern leaf blight), *Venturia inaequalis* (apple scab).
  Deuteromycotina (sexual stage absent or unknown)
    Coleomycetes—*Ascochyta pisi* (pea blight), *Phoma lingam* (crucifer black leg), *Septoria nodorum* (wheat glume blotch), *S. tritici* (wheat leaf blotch).
    Hyphomycetes—*Alternaria* spp. (leaf spots and blights). *Botrytis cinerea* (grey mould), *Cercospora* spp. (leaf spots), *Fulvia fulva* (tomato leaf mould), *Fusarium* spp. (foot and root rots) and *F. oxysporum* (vascular wilts), *Pseudocercosporella herpotrichoides* (cereal eye spot). *Rhynchosporium secalis* (barley leaf blotch), *Verticillium albo-atrum* and *V. dahliae* (vascular wilts).
  Basidiomycotina (sexual spores = Basidiospores)
    Hemibasidiomycetes—Smuts, e.g. *Ustilago* spp. (e.g. *U. maydis,* maize smut). Rusts, e.g. *Puccinia* (e.g. *P. graminis,* cereal black stem rust; *P. striiformis,* cereal yellow stripe rust), *Melampsora lini* (flax rust), *Hemileia vastatrix* (coffee rust), *Uromyces* spp. (e.g. *U. appendiculatus,* bean rust).
    Hymenomycetes—*Armillariella mella* (root rot/honey fungus), *Chondrostereum purpureum* (plum/cherry silver leaf), *Stereum* spp. (heart rots), *Thanatephorus cucumeris (= Rhizoctonia solani)* (root and stem rots), *Athelia rolfsii (= Sclerotium rolfsii)* (stem and root rots).

[a] Only subdivisions and classes with pathogens of major importance are included.

Myxomycota are classified as fungi, the phylogenic relationships between the two divisions are unclear. The Myxomycota contains few plant pathogens of importance with the notable exception of the *Brassica* club-root fungus *Plasmodiophora brassicae* (4). *Polymyxa graminis* and *P. betae* are, however, important as vectors of plant viruses (5).

1.3 Hyphal structure

Although several plant pathogenic fungi are capable of yeast-like budding growth *in vivo* (e.g. vascular phase of *Fusarium oxysporum*, *Verticillium*, and *Ophiostoma* wilts) or *in vitro* (e.g. smuts), the vast majority have some form of mycelial organization composed of branched filaments or hyphae (6). Fungi in the Mastigomycotina and Zygomycotina are generally coenocytic, that is their hyphae lack septa. The hyphae of Ascomycotina and Deuteromycotina are normally septate but have pores through which nuclei and mitochondria may pass. In many of the Basidiomycotina the heterokaryotic secondary mycelium is septate but the septum is more complex in fine structure, and is known as the dolipore septum. Smuts and rusts do not have dolipore septa.

1.4 Cell wall composition

The main components of fungal cell walls are carbohydrates in the form of polysaccharides or glycoproteins. The main structural components of the wall consist of microfibrils of β-1,4-N-acetylglucosamine (chitin), β-1,4-glucan (cellulose), or β-1,3 and β-1,6 linked glucans (7, 8, 9). The occurrence of these polymers differs between taxonomic groups. Chitin is the most common structural polymer and has been found in all groups, except the oomycetes where cellulose is found, and hemiascomycetes which contain β-1,3 and β-1,6 linked glucans. The cellulose content of cell walls, L-lysine biosynthetic pathway, and rRNA sequence data suggests a separate evolutionary origin for the oomycetes, possibly with a closer phylogenetic relationship to the brown algae (10). A wide variety of matrix polymers are found including glucans, mannans, and β-1,4 glucosamine (chitosan). The fungal cell wall is likely to play a critical role in host–parasite recognition and cell wall components have been shown to act as elicitors of host defence responses (11, 12).

2. Genome organization

Filamentous fungi typically have a haploid genome size of around 2–5×10^7 base pairs organized as linear nuclear chromosomes associated with histones. Structural genes involved in the same metabolic pathway are usually unlinked, but there are several exceptions. These include the quinic acid utilization gene clusters of *Neurospora crassa* (13) and *Aspergillus nidulans* (14) and the nitrate (15) and proline catabolism (16) gene clusters of *A. nidulans*. Examples also exist of 'cluster genes' which encode multifunctional polypeptides such as the *arom*A gene of *A. nidulans* (17). Studies of nuclear, protein encoding genes have revealed a number of key features including putative 5′ regulatory sequences (TATAAA and CCAAT), transcription start points (tsp), introns, and 3′ polyadenylation sites (18, 19). Genes have been cloned from a taxonomically wide range of fungal plant pathogens including oomycetes, pyrenomycetes, and teliomycetes (see *Table 2*) and as sequence information accumulates, it will be interesting to see if novel structural or regulatory features emerge. The *hsp*70 gene from *Bremia lactucae* (oomycetes) consists of a single open reading frame (ORF) with multiple CCAAT motifs and an A + T rich putative 'TATA' box upstream (5′) from the tsp and a canonical eukaryotic polyadenylation recognition sequence (AATAAA)

Table 2. Genes clones from fungal plant pathogens

Species	Gene	System	Ref.
Bremia lactucae	*hsp*70	heat shock[d]	(20)
Cochliobolus carbonum	*pgn*1	endopolygalacturonase[i]	(23)
Cochliobolus heterostrophus	*trp*1	phosphoribosylanthranilate isomerase[c]	(24)
		17S, 5.85 and 25S rRNA genes[i]	(25)
Colletotrichum graminicola	*tub*1 *tub*2	β-tubulin[d]	(26)
Cryphonectria parasitica	gpd-1	glyceraldehyde 3 phosphate dehydrogenase[d]	(27)
Erysiphe graminis	*tub*B	β-tubulin	(28)
Fulvia fulva		reverse transcriptase[e]	(29)
Fusarium sporotrichiodes	*tox*5	trichodiene synthase[e]	(30)
Nectria haematococca	*pda*1	pisatin demethylase[g]	(31)
	*cut*A	cutinase[e]	(21)
Sclerotinia sclerotiorum		β-glucosidase[h]	(32)
Septoria nodorum	*tub*A[R]	β-tubulin[d]	(33)
Uromyces appendiculatus	*inf*24	infection structure specific[f]	(22)
Ustilago maydis	*rec*1	DNA repair recombination[a]	(34)
		ornithine N^5 oxygenase[a]	(35)
	*pyr*3	dihydroorotase[c]	(36)
	*acu*A	acetyl-CoA[d]	(37)
	*pyr*6	orotidine decarboxylase[b]	(38)
	*b*1, *b*2	mating type[a]	(39)
	leu 1	leucine biosynthesis[a]	(40)
	ums 1	heat shock[d]	(41)
	*ums*2, *ums*3	heat shock	
	*ums*4	heat shock	
	gpd	glyceraldehyde-3-phosphate dehydrogenase[d]	(42)

Cloning strategy
[a] direct complementation
[b] complementation of *S. cerevisiae* mutant
[c] complementation of *E. coli* mutant
[d] heterologous probe
[e] antibody probe
[f] differential hybridization
[g] transformation of *A. nidulans* to *Pda* +
[h] expression in *E. coli*
[i] differential centrifugation
[j] oligonucleotide probe

downstream (3′) from the coding region (20). The cutinase gene of *Nectria haematococca* (pyrenomycetes) contains one short 51 bp intron, and A + T rich sequence upstream from the tsp and also a downstream sequence similar to the polyadenylation site (21). An unusual feature of the *inf* 24 gene of the rust *Uromyces appendiculatus* (teliomycete) is the presence of long untranslated regions 5′ and 3′ of the longest transcript including two small ORFs prior to the ATG codon of the largest ORF (22). A putative 'TATA' box was identified 23 nucleotides upstream from the tsp of the longest ORF. The glyceraldehyde-3-phosphate dehydrogenase gene of *Ustilago maydis* is unusual in containing a relatively large 407 bp intron (42).

3. Genetic analysis

Although the frequency and importance of sexual recombination in wild populations of many fungi is largely unknown, genetic analysis through the sexual cycle is applicable to a wide range of fungi (for recent reviews see ref. 43). The majority of plant pathogens for which sexual analysis has been developed are heterothallic, i.e. two compatible strains of different mating type are required to produce sexual spores, and therefore the sexual spores should all be of hybrid origin. Naturally occurring genetic markers have been crucial to the development of genetic analysis for biotrophic pathogens and detailed analyses of specific virulence characteristics have been carried out for several foliar biotrophs (see Section 4). Other naturally occurring markers include fungicide resistance, isozymes, and restriction fragment length polymorphisms (RFLPs) (44, 45). An advantage of isozymes and RFLPs is that they are generally co-dominant and the genotype can be determined directly from the electrophoretic phenotype. Isozymes have proved valuable markers in the genetic analysis of the vegetatively diploid fungus *P. infestans* (46, 47). For vegetatively haploid species which can be cultured axenically, auxotrophic markers can usually be readily isolated following ultraviolet or chemical mutagenesis (e.g. using nitrosoguanidine or ethyl methane sulphonate). In addition to the analysis of host species and cultivar specificity, sexual crosses have also been used to study aggressiveness (e.g. ref. 48) and several specific biochemical traits associated with pathogenicity (see Section 5). Many important plant pathogens have no known sexual stage (classified in the Deuteromycotina, *Table 1*), but genetic analysis may still be possible using the parasexual cycle (49, 50). The essential features are the construction of heterozygous diploids between genetically marked haploids by hyphal or protoplast fusion, followed by mitotic crossing over and spontaneous or induced haploidization. Using parasexual analysis, Typas and Heale (51) mapped 33 auxotrophic and morphological markers to 5 linkage groups in the vascular wilt fungus *Verticillium*.

4. Genetics of the host–pathogen interaction

Our understanding of the genetics of the interaction between plants and their microbial pathogens stems from the research carried out by Flor (52, 53) in which he studied simultaneously the genetic basis of host resistance and pathogen virulence in the flax rust (*Linum usitatissimum/Melampsora lini*) interaction. Flor demonstrated that both host and pathogen appeared to have evolved complementary genetic systems such that for each gene conditioning rust reaction in the host there was a gene conditioning pathogenicity in the parasite (54). Because rust fungi are dikaryotic during the major part of the disease cycle it was also possible to study dominance relationships. Since the pioneering experiments of Flor, other plant–pathogen relationships have been shown to follow such a 'gene-for-gene' interaction (55, 56). Most of these studies have been with obligate biotrophic fungi and the best characterized interactions are flax rust (57, 58) and lettuce downy mildew (59). Gene-for-gene interactions have been suggested for many other diseases where pathogen races can be identified by the differential reactions of host cultivars with known genotypes, but where genetic analysis of the pathogen is lacking. Several generalizations emerge from these genetic studies: host resistance and pathogen avirulence are usually dominant, and linkage between host resistance genes is quite common but linkage between avirulence genes is not. However, exceptions to these typical characteristics do occur and have been discussed by Crute (56) and Gilchrist *et al.* (60). These include recessive or incompletely dominant resistance or avirulence alleles, the presence of genes which modify the expression of resistance or avirulence alleles and apparent departures from a strict 1 : 1 numerical ratio of host and pathogen genes.

The genetic evidence has focused attention on the possible biochemical mechanisms underlying specificity between host cultivars and pathogen races. Ellingboe (61) suggested that the gene products of dominant resistance and avirulence alleles interact directly and specifically to form a dimer resulting in a resistant or incompatible interaction. Lack of specific interaction or 'recognition' leads to a susceptible or compatible interaction. Other models (62, 63) do not preclude indirect interaction between resistance and avirulence alleles; for example Albersheim and Anderson-Prouty (56) envisaged avirulence genes coding for glycosyl transferases responsible for determining the carbohydrate components of surface glycoproteins involved in recognition. Specific recognition between host and pathogen then leads to the coordinated activation of host-defence genes such as those involved in phytoalexin biosynthesis, cell wall modifications, or those coding for extracellular hydrolytic enzymes.

Several avirulence genes from bacterial plant pathogens have now been cloned and sequenced (65, 66). These studies indicate that *avr* genes code for cytoplasmic proteins but the biochemical function of these genes has not been resolved. The large genome size, generally low frequencies of transformation, and obligate biotrophic nature of many of the well-characterized fungal pathogens suggest that alternative strategies such as RFLP mapping may be necessary for the cloning of fungal avirulence genes. In the absence of this molecular evidence, alternative and/or multiple explanations of specificity in fungi cannot be discounted.

5. Fungal pathogenicity determinants

5.1 Polymer degrading enzymes

The aerial parts of plants are protected by the cuticle of which the major structural component is a complex insoluble polyester called cutin. Many fungi penetrate this layer directly and there is very good evidence for the involvement of the enzyme cutinase in the penetration of plant cuticle by *Nectria haematococca* (67, 68). Electron microscopy studies using ferritin-labelled antibodies prepared against cutinase have confirmed that *N. haematococca* secretes cutinase during penetration of the cuticle. Specific antibodies or a chemical cutinase inhibitor prevented infection and a correlation exists between the level of cutinase associated with germinating spores and the ability of different strains of *N. haematococca* to infect pea stems. Cutinase cDNA (69) and genomic clones (21) have been isolated and sequenced. Some fungi evade the cuticular barrier, for example by penetration via stomata or damaged plant surfaces. Subsequent colonization however, usually involves repeated penetration of cell walls and extracellular enzymes which can degrade the major components of plant cell walls are produced by the majority of plant pathogenic fungi. There is good biochemical evidence that enzymes capable of solubilizing the pectic polysaccharides of the middle lamella and primary cell wall (pectic enzymes) are important pathogenicity determinants (70), but much of the evidence to implicate other depolymerases is largely circumstantial. Genetic approaches involving the isolation of enzyme deficient mutants are in many cases complicated by multiple isozymes and the requirement for detection media specific for the numerous possible forms, e.g. pectin or pectate hydrolases or lyases. Where multiple isozymes exist, structural gene mutations may be difficult to identify as the majority of mutants with an enzyme deficient phenotype may be blocked at regulatory or secretory loci (71). Gene cloning opens the possibility in some fungi of creating site-specific mutations by transformation and gene disruption, thereby giving an unequivocal answer to the role of fungal depolymerases. This has been achieved recently for *Cochliobolus carbonum* a pathogen of maize, where gene disruption has been used to demonstrate that endopolygalacturonase is not required for pathogenicity on this host (23).

5.2 Phytoalexin detoxifying enzymes

The ability of *N. haematococca* to demethylate pisatin is required for its tolerance to this phytoalexin and for pathogenicity towards pea (72). Genetic analyses have identified several loci which quantitively determine the level of pisatin demethylase activity (73) and the *pda*1 gene has been cloned (31). A recent study of the potato pathogen *Gibberella pulicaris* (74) has shown analogously that tolerance to rishitin and high virulence co-segregated in genetic crosses, suggesting that a high level of phytoalexin tolerance is either required or closely linked to genes for virulence.

5.3 Toxins

Toxins can be defined as microbial products which are directly damaging to plant cells and which are involved in disease development. Fungal toxins are a diverse group of low molecular weight secondary metabolites (75). A number of fungi, including several species of *Cochliobolus* (anamorph: *Helminthosporium*) and *formae speciales* of *Alternaria alternata* produce toxins in axenic culture which show similar host specificity to the fungus (host selective toxins). Many non-host selective toxins have also been characterized including fusicoccin produced by *Fusicoccum amygdali* and fusaric acid by *Fusarium oxysporum* (76). Genetic studies of toxin production in *C. carbonum* (HC toxin), *C. victoriae* (HV toxin), and *C. heterostrophus* (T. toxin) have shown that in each species a single locus determined toxin production and that host or cultivar specificity segregates with toxin production (77, 78). Turgeon *et al.* (79) have recently described a molecular strategy for the isolation of the *TOXI* allele of *C. heterostrophus* race-T which involves transformation of a race-O strain with a cosmid bank of wild type race-T DNA. Selection for transformants containing *TOXI* utilizes a microbiological assay based on the use of a recombinant *E. coli* strain containing the maize T cytoplasm mitochondrial gene *urf*13-*T*. It is not known how single genetic loci control the biosynthesis of these toxins although it is possible that they are regulatory, encode a multifunctional peptide, or represent a tightly-linked gene cluster. The recent purification of two separate enzymes involved in the biosynthesis of HC toxin, the cyclic peptide produced by *C. carbonum* favours the latter explanation for this fungus and should facilitate the cloning of these two genes (80).

5.4 Plant growth regulators

There is convincing biochemical and molecular evidence for the involvement of auxins and cytokinins in the diseases caused by the plant pathogenic bacteria *Agrobacterium tumefaciens*, *A. rhizogenes*, and *Pseudomonas savastanoi* (81). Although the disease symptoms produced by a wide range of fungi such as *Ustilago maydis* and *Plasmodiophora brassicae* suggest the involvement of plant growth regulators and the fact that many fungi can synthesize these *in vitro*, a conclusive role in pathogenesis has not been established (82).

6. Obtaining cultures of fungal plant pathogens

Fungal cultures may be obtained directly from diseased plant material, from other mycologists/plant pathologists, University or Research Institute culture collections, or National collections. The Plant Pathologists Pocketbook (83) lists the addresses of National Culture collections, and is a valuable source of information on the isolation, culture, and preservation of plant pathogens. The addresses of the UK and USA collections are:

The Culture Collection,
CAB International,
Mycological Institute,
Ferry Lane,
Kew, Surrey, TW9 3AF,
UK.

American Type Culture Collection,
12301 Parklawn Drive,
Rockville,
MD 20852,
USA.

References

1. Lewis, D. H. (1973). *Biol. Rev. Cambridge Philosophic Soc.*, **48**, 261.
2. Maclean, D. J. (1982). In *The rust fungi* (ed. K. J. Scott and A. K. Chakravorty), p. 37. Academic Press, London.
3. Ainsworth, G. C. (1973). In *The fungi: an advanced treatise*, Vol. 4B (ed. G. C. Ainsworth, F. K. Sparrow, and A. S. Sussman), p. 1. Academic Press, London.
4. Buczacki, S. T. (1983). In *Zoosporic plant pathogens—a modern perspective* (ed. S. T. Buczacki), p. 161. Academic Press, London.
5. Teakle, D. S. (1983). In *Zoosporic plant pathogens—a modern perspective* (ed. S. T. Buczacki), p. 233. Academic Press, London.
6. Moore, R. T. (1965). In *The fungi: an advanced treatise*, Vol. 1 (ed. G. C. Ainsworth and A. S. Sussman), p. 95. Academic Press, London.
7. Bartnicki-Garcia, S. (1968). *Annu. Rev. Microbiol.*, **22**, 87.
8. Bartnicki-Garcia, S. (1970). In *Phytochemical phylogeny* (ed. J. B. Marborne), p. 81. Academic Press, London.
9. Peberdy, J. F. (1990). In *Biochemistry of cell walls and membranes in fungi* (ed. P. J. Kuhn, A. P. J. Trinci, M. J. Jung, M. W. Goosey, and L. G. Copping), p. 5. Springer, Berlin.
10. Gunderson, J. H., Elwood, H., Ingold, A., Kindle, K., and Sogin, M. L. (1987). *Proc. Natl. Acad. Sci. USA*, **81**, 5724.
11. Loschke, D. C., Hadwiger, L. A., and Wagoner, W. (1983). *Physiol. Plant Pathol.*, **23**, 163.
12. Ryder, T. B., Cramer, C. L., Bell, J. N., Robbins, M. P., Dixon, R. A., and Lamb, C. J. (1984). *Proc. Natl. Acad. Sci. USA*, **81**, 5724.
13. Geever, R. F., Hueit, L., Baum, J. A., Tyler, B. M., Patel, V. B., Rutledge B, J., *et al.* (1989). *J. Mol. Biol.*, **207**, 15.
14. Lamb, H. K., Hawkins, A. R., Smith, M., Harvey, I. J., Brown, J., Turner, G., *et al.* (1990). *Molec. Gen. Genet.*, **223**, 17.
15. Johnstone, I. L., McCabe, P. C., Greaves, P., Gurr, S. J., Cole, G. E., Brow, M. A. D., *et al.* (1990). *Gene*, **90**, 181.
16. Arst, H. N., Jr. and Scazzocchio, C. (1985). In *Gene manipulations in fungi* (ed. J. W. Bennett and L. L. Lasure), p. 309. Academic Press, London.
17. Hawkins, A. R. (1987). *Curr. Genet.*, **11**, 491.
18. Gurr, S. J., Unkles, S. E., and Kinghorn, J. R. (1987). In *Gene structure in eukaryotic microbes* (ed. J. R. Kinghorn), p. 93. IRL, Oxford.
19. Ballance, D. J. (1990). In *Molecular industrial mycology* (ed. S. A. Leong and R. M. Berka), p. 1. Marcel Dekker, New York.
20. Judelson, H. S. and Michelmore, R. W. (1989). *Gene*, **79**, 207.
21. Soliday, C. L., Dickman, M. B., and Kolatukkudy, P. E. (1989). *J. Bacteriol.*, **171**, 1942.
22. Bhairi, S. M., Staples, R. C., Freve, P., and Yoder, O. C. (1989). *Gene*, **81**, 237.
23. Scott-Craig, J. S., Panaccione, D. G., Cervone, F., and Walton, J. D. (1990). *The Plant Cell*, **2**, 1191.
24. Turgeon, B. G., MacRae, W. D., Garber, R. C., Fink, G. R., and Yoder, O. C. (1986). *Gene*, **42**, 79.
25. Garber, R. C., Turgeon, B. G., Selker, E. U., and Yoder, O. C. (1988). *Curr. Genet.*, **14**, 573.

26. Panaccione, D. G. and Hanaie, R. M. (1990). *Gene*, **86**, 163.
27. Choi, G. H. and Nuss, D. L. (1990). *Nucleic Acids Res.*, **18**, 5566.
28. Sherwood, J. E. and Sommerville, S. C. (1990). *Nucleic Acids Res.*, **18**, 1052.
29. McHale, M. T., Roberts, I. N., Talbot, N. J., and Oliver, R. P. (1989). *Molec. Plant–Microbe Interactions*, **2**, 165.
30. Hohn, T. M. and Beremand, P. D. (1989). *Gene*, **79**, 131.
31. Weltring, K. M., Turgeon, B. G., Yoder, O. C., and VanEtten, H. D. (1988). *Gene*, **68**, 335.
32. Waksman, G. (1989). *Curr. Genet.*, **15**, 295.
33. Cooley, R. N. and Caten, C. E. (1989). In *Proceedings of the EMBO–Alko workshop on the molecular biology of filamentous fungi*, Helsinki, 1989 (ed. H. Nevalainen and M. Pentila). Foundation for Biotechnical and Industrial Fermentation Research, **6**.
34. Holden, D. W., Spanos, A., and Banks, G. R. (1989). *Nucleic Acids Res.*, **17**, 10489.
35. Wang, J., Buddle, A. D., and Leong, S. A. (1989). *J. Bacteriol.*, **171**, 2811.
36. Banks, G. R. and Taylor, S. Y. (1988). *Mol. Cell Biol.*, **8**, 5417.
37. Hargreaves, J. A. and Turner, G. (1989). *J. Gen. Microbiol.*, **135**, 2675.
38. Kronstad, J. W., Wang, J., Covert, S. F., Holden, D. W., McKnight, G. L., and Leong, S. A. (1989). *Gene*, **79**, 97.
39. Kronstad, J. W. and Leong, S. A. (1989). *Proc. Natl. Acad. Sci. USA*, **86**, 978.
40. Fotheringham, S. and Holloman, W. K. (1989). *Mol. Cell Biol.*, **9**, 4052.
41. Holden, D. W., Kronstad, J. W., and Leong, S. A. (1989). *EMBO J.*, **8**, 1927.
42. Smith, T. L. and Leong, S. A. (1990). *Gene*, **93**, 111.
43. Sidhu, G. S. (1988). *Advances in Plant Pathology*, **6**, 33.
44. Newton, A. C. (1987). In *Genetics and plant pathogenesis* (ed. P. R. Day and G. J. Jellis), p. 187. Blackwell, Oxford.
45. Michelmore, R. W. and Hulbert, S. H. (1987). *Annu. Rev. Phytopathol.*, **25**, 383.
46. Shattock, R. C., Tooley, P. W., and Fry, W. E. (1986). *Phytopathology*, **76**, 410.
47. Shattock, R. C., Tooley, P. W., Sweigard, J., and Fry, W. E. (1987). In *Genetics and plant pathogenesis* (ed. P. R. Day and G. J. Jellis), p. 175. Blackwell, Oxford.
48. Caten, C. E. *et al.* (1984). *Can. J. Bot.*, **62**, 1209.
49. Pontecorvo, G. (1956). *Annu. Rev. Microbiol.* **1**, 393.
50. Tinline, R. D. and MacNeil, G. B. (1969). *Annu. Rev. Phytopathol.*, **7**, 147.
51. Typas, M. A. and Heale, J. B. (1978). *Genet. Res. Camb.*, **31**, 131.
52. Flor, H. H. (1946). *J. Agric. Res.*, **73**, 335.
53. Flor, H. H. (1947). *J. Agric. Res.*, **74**, 241.
54. Flor, H. H. (1956). *Adv. Genet.*, **8**, 29.
55. Day, P. R. (1974). *Genetics of host parasite interaction*. Freeman, San Francisco.
56. Crute, I. R. (1985). In *Mechanisms of resistance to plant diseases* (ed. R. S. Fraser), p. 80. Martinus Nijhoff and W. Junk, Dordrecht.
57. Lawrence, G. J., Mayo, G. M. E., and Sheperd, K. W. (1981). *Phytopathology*, **71**, 12.
58. Lawrence, G. J., Shepherd, K. W., and Mayo, G. M. E. (1981). *Heredity*, **46**, 297.
59. Crute, I. R. (1987). In *Genetics and plant pathogenesis* (ed. P. R. Day and G. J. Jellis), p. 207. Blackwell, Oxford.
60. Gilchrist, B. J., Person, C. O., and Pope, D. D. (1987). In *Populations of plant pathogens: their dynamics and genetics* (ed. M. S. Wolfe and C. E. Caten), p. 7. Blackwell, Oxford.
61. Ellingboe, A. H. (1982). In *Active defense mechanisms in plants* (ed. R. K. S. Wood), p. 179. Plenum, New York.

62. Keen, N. T. (1985). In *Genetic basis of biochemical mechanisms of plant disease* (ed. J. V. Groth and W. R. Bushnell), p. 85. A.P.S. Press, St Paul, MN.
63. Gabriel, D. N. and Rolfe, B. G. (1990). *Annu. Rev. Phytopathol.*, **28**, 365.
64. Albersheim, P. and Anderson-Prouty, A. J. (1975). *Annu. Rev. Plant Physiol.*, **26**, 31.
65. Napoli, C. and Staskawicz, B. J. (1987). *J. Bacteriol.*, **169**, 572.
66. Ronald, P. C. and Staskawicz, B. J. (1988). *Molec. Plant–Microbe Interactions*, **1**, 191.
67. Kolatukkudy, P. E. (1985). *Annu. Rev. Phytopathol.*, **23**, 223.
68. Kolatukkudy, P. E. and Crawford, M. S. (1987). In *Molecular determinants of plant diseases* (ed. S. Nishimura), p. 75. Japanese Science Society Press/Springer, Berlin.
69. Soliday, C. L., Flurkey, W. H., Okita, T. W., and Kolatukkudy, P. E. (1984). *Proc. Natl. Acad. Sci. USA*, **81**, 3939.
70. Cooper, R. M. (1984). In *Biochemical plant pathology* (ed. J. A. Callow), p. 101. Wiley, Chichester and New York.
71. Cooper, R. M. (1987). In *Genetics and plant pathogenesis* (ed. P. R. Day and G. J. Jellis), p. 261. Blackwell, Oxford.
72. VanEtten, H. D., Matthews, D. E., and Matthews, P. S. (1989). *Annu. Rev. Phytopathol.*, **27**, 143.
73. VanEtten, H. D. and Kistler, H. C. (1988). *Advances in Plant Pathology*, **6**, 189.
74. Desjardins, A. E. and Gardiner, H. W. (1989). *Molec. Plant–Microbe Interactions*, **2**, 26.
75. Scheffer, R. P. and Livingston, R. S. (1984). *Science*, **223**, 17.
76. Stoessl, A. (1981). In *Toxins in plant disease* (ed. R. D. Durbin), p. 109. Academic Press, London.
77. Scheffer, R. P., Nelson, R. R., and Ullstrup, A. J. (1967). *Phytopathology*, **57**, 1288.
78. Yoder, O. C. and Gracen, V. E. (1975). *Phytopathology*, **65**, 273.
79. Turgeon, B. G., Cuiffetti, L., Schafer, W., and Yoder, O. C. (1989). In *Phytotoxins and plant pathogenesis* (ed. A. Graniti, R. D. Durbin, and A. Ballio), p. 409. Springer, Berlin.
80. Walton, J. D. and Holden, F. R. (1988). *Molec. Plant–Microbe Interactions*, **1**, 128.
81. Morris, R. O. (1986). *Annu. Rev. Plant Physiol.* **37**, 509.
82. Pegg, G. F. (1984). In *Plant diseases: infection, damage and loss* (ed. R. K. S. Wood and G. J. Jellis), p. 29. Blackwell, Oxford.
83. Commonwealth Agricultural Bureau (1983). *Plant pathologists pocketbook*, 2nd edn. CAB, [Town].

8

Gene transformation in plant pathogenic fungi

JOHN HARGREAVES and GEOFFREY TURNER

1. Introduction

Gene transformation of filamentous fungi is a relatively new but rapidly developing area (1). The first report of DNA-mediated transformation of a filamentous fungus was with *Neurospora crassa* (2) and was soon followed by the demonstration of gene transfer in another ascomycete, *Aspergillus nidulans* (3, 4). Since then a number of genes have been isolated by expression from these and other fungi, and molecular genetic techniques are now being extended to a wide range of commercially and agriculturally important fungal species. This chapter aims to provide a guide for the development of transformation systems for plant pathogenic fungi. Emphasis is placed on (a) choice of genes as selectable markers, (b) development of cloning vectors, and (c) procedures used to make fungal cells permeable to large DNA molecules.

Transformation vectors and protocols vary from fungus to fungus and from laboratory to laboratory and thus only two detailed protocols are provided. The first, which is efficient, reliable, and yields high numbers of transformants (*c.* 10^4 transformants/μg DNA) is for the maize smut pathogen, *Ustilago maydis*. The second is for the bean anthracnose pathogen, *Glomerella cingulata f. sp. phaseoli* (anamorph: *Colletotrichum lindemuthianum*), where transformation is often erratic and transformant numbers are much lower than with *U. maydis* (1–50/μg DNA). The latter protocol is included because it is representative of transformations with most other pathogenic fungi. The isolation of pathogenicity genes by expression is discussed briefly in Section 5.

2. Transformation vectors

2.1 Selectable markers for transformation of plant pathogens

Table 1 lists selectable marker genes used for transformation of plant pathogens. While fungi of most phylogenetic classes are represented, oomycetes are conspicuous by their absence. Genes employed as selectable markers for transformation fall into two main groups:

Table 1. Selectable genes for transforming plant pathogenic fungi.

Pathogen	Transformation frequency (number of transformants/μg)	Selectable marker	Reference
Botryotinia squamosa[a]	0.2–1	*hygB*[R5]	(5)
Cercospora kikuchii	1–2	*ben*[R1]	(6)
Claviceps purpurea	0.2–0.3	*bleo*[R5]	(7)
Cochliobolus heterostrophus	0.02–0.6	*amdS*[+]	(9)
	0.2–10	*hygB*[R3]	(9)
Colletotrichum capsici	c. 20*	*hygB*[R8]	(10)
C. graminicola	13–28	*ben*[R1.2]	(11)
C. trifolii	10–20*	*hygB*[R3]	(12)
	0.4–1*	*ben*[R2]	(12)
Cryphonectria parasitica[b]	>10^5	*hygB*[R5]	(13)
	1–14	*ben*[R2]	(13)
Fulvia fulva	c. 10	*hygB*[R4]	(14)
Fusarium moniliforme[c]	nk	*hygB*[R]	(15)
F. oxysporum	1.0	*hygB*[R5]	(16)
	10	*niaD*[2]	(17)
F. sambucinum[d]	nk	*hygB*[R]	(18)
F. solani f. sp. pisi	c. 10*	*hygB*[R8]	(10)
F. solani f. sp. phaseoli	3.3*	*aph(3')II*[g]	(19)
Glomerella cingulata	0.2	*amdS*[+]	(20)
f. sp. phaseoli[e]	0.2	*amdS*[+]	(20)
	0.2	*hygB*[R3]	(20)
	2–50	*hygB*[R4]	(unpublished)
	us	*hygB*[R10]	(unpublished)
	5	*niaD*[2]	(21)
Gaeumannomyces graminis	10	*ben*[R2]	(22)
Leptosphaeria maculans	60	*hygB*[R4]	(23)
	nk	*ben*[R2]	(24)
Magnaporthe grisea[f]	c. 35	*argB*[2]	(25)
	nk	*hygB*[R4]	(26)
	4.4	*niaD*[2]	(21)
Nectria haematococca[g]	c. 1	*niaD*[2]	(21)
Ophiostoma ulmi[h]	us	*ben*[R2]	(27)
	nk	*hygB*[R6]	(27)
Pseudocercosporella herpotrichoides	1–20	*ben*[R1]	(28)
	1–20	*hygB*[R]	(28)
Septoria nodorum	2–20	*hygB*[R4]	(29)
	nk	*ben*[R1.2]	(24)
Ustilago maydis	50**	*hygB*[R7]	(30)
	1000	*hygB*[R7]	(30)
	>10^4***	*hygB*[R7]	(31)
	us	*hygB*[R5]	(32)
	us	*hygB*[R4]	(unpublished)
	10–15	*neo*[R]	(33)
	40–90	*pyr3*[1]	(34)
	5–20	*acuA*[1]	(35)
U. violacea	60–80*	*hygB*[R3]	(36)

us transformation unsuccessful, nk transformation frequency not known, (a) anamorph *Botrytis squamosa*, (b) anamorph *Endothia parasitica*, (c) anamorph *Gibberella fujikuroi*, (d) anamorph *Gibberella pulicaris*, (e) anamorph *Colletotrichum lindemuthianum*, (f) anamorph *Pyricularia oryzae*, (g) anamorph *Fusarium solani*, (h) anamorph *Ceratocystis ulmi*, * transformed using the lithium procedure, ** transformed with linear

(a) Those which confer dominant or semi-dominant drug (antibiotic or fungicide) resistance in normally sensitive species or strains.

(b) Those which complement nutritional (usually auxotrophic) mutations, or which allow the fungus to use a substrate that is not normally utilized.

Of the markers used, antibiotic and fungicide resistance genes have found the greatest application due to the ease with which their expression can be detected without the need to isolate specific mutants. Higher transformation frequencies have often been achieved when selection is based on complementation of a nutritional mutant, but there have been few attempts to transform plant pathogens in this way.

2.1.1 Antibiotic and fungicide resistance genes

Selection of transformants by enhanced resistance to drugs is usually based on the following genes.

(a) *ben*^{R} genes. These encode forms of β-tubulin that are insensitive to the systemic fungicide benomyl (methyl 1-(butylcarbamoyl) benzimidazolecarbamate, Du Pont de Nemours & Co.) or to the related compound carbendazim. Gene transfer based on resistance to benomyl was first demonstrated with *N. crassa* (40) and both the *N. crassa ben*^{R} gene and homologous resistance genes (i.e. genes from the fungal species being transformed) have since been used to transform a number of different plant pathogens (*Table 1*). Unfortunately, transformation frequencies obtained with these fungi are much lower than those achieved with *N. crassa* (15 000 transformants/μg DNA). This suggests that, in many fungal species, the resistance conferred by *ben*^{R} genes is insufficient for a primary selection. Furthermore, the *ben*^{R} genes may exhibit only partial dominance over the wild-type gene in which case the relative levels of expression of the resistant β-tubulin gene may limit the success of this transformation system. Nevertheless, a wide range of fungi are sensitive to the benzimidazole fungicides within the range 0.1–2.0 μg/ml.

(b) *hygB*^{R} genes. These encode the enzyme hygromycin B phosphotransferase, which phosphorylates the broad spectrum antibiotic hygromycin B, rendering it inactive. *HygB*^{R} genes are found on the *Escherichia coli* plasmid pKC203 and in

plasmid, *** transformed with a plasmid containing an autonomous replicating sequence (ARS) from *U. maydis*.

acuA acetyl-CoA synthase gene of *U. maydis*, *amdS* $^+$ acetamidase gene of *A. nidulans*, *aph*(3')II G418 resistance gene from transposon Tn5, *argB* ornithine carbamoyltransferase gene from *A. nidulans*, *ben*^{R} benomyl resistance gene, *bleo*^{R} (*phleo*^{R}) bleomycin (phleomycin) resistance gene from transposon Tn5, *hygB*^{R} hygromycin B phosphotransferase gene, *neo*^{R} aminoglycoside phosphotransferase, *niaD* nitrate reductase gene from *A. nidulans*, *pyr*3 dihydroorotase gene from *U. maydis*.

1 transformation with a homologous selectable gene, 2 transformation with a heterologous selectable gene, 3 gene fused to promoter elements of *C. heterostrophus* (9), 4 gene fused to the *A. nidulans gpd*A promoter and *trp*C terminator sequences (37), 5 gene fused to promoter sequences of the *trp*C gene from *A. nidulans* (32), 6 gene fused to the promoter for isopenicillin N synthase from *Penicillium chrysogenum* (27), 7 gene fused to the *hsp*70 gene from *U. maydis* (30), 8 gene fused to the cutinase promoter from *F. solani f. sp. pisi* (10), 9 gene fused to the 35S promoter from cauliflower mosaic virus (19), 10 gene fused to the *cyc*1 gene from *Saccharomyces cerevisiae* (38).

Streptomyces hygroscopicus. It is necessary to place the $hygB^R$ coding sequence under the control of a promoter which functions in the fungus to be transformed before selection on hygromycin. Several suitable vectors have been constructed (see *Table 2*). When selecting a $hygB^R$ based vector, it should be borne in mind that a promoter from one class of fungi may not function in a fungus from another class. For example, the plasmids pDH25 and pAn7-1, in which the $hygB^R$ gene is under the control of an *A. nidulans* promoter, do not transform the hemibasidio-mycete fungus, *U. maydis*. Similarly, promoters from *U. maydis* and *Saccharomyces cerevisiae* do not function in some euascomycete fungi (see Table 1). There are, however, a number of exceptions where a promoter will operate in fungi from another class (33, 36). If no suitable gene constructions are available it will be necessary to place the $hygB^R$ gene downstream of a 'strong' homologous promoter sequence by making a gene fusion (30) or by linking the $hygB^R$ gene to a known promoter sequence (37, 41). Alternatively, a suitable promoter may be isolated, at random, as has been done from the maize pathogen *Cocchliobolus heterostrophus* (9). Unfortunately, some fungi and even strains of the same species are not sufficiently sensitive to hygromycin B to permit this selection. Concentrations ranging from 50 to 200 μg/ml are normally used to inhibit growth.

(c) *bleo*^{R} or *phleo*^{R} genes. These confer resistance to bleomycin and the related antibiotic, phleomycin respectively and are derived from three different sources: the transposon Tn5, a staphylococcal plasmid (pUB110) and *Streptoalloteichus hindustanus*. As with the $hygB^R$ genes, they must be placed under the control of a suitable fungal promoter. Typical concentrations required to inhibit growth range from 20 to 50 μg/ml.

(d) *oliC*^{R} gene. This encodes a form of subunit 9 of the mitochondrial ATP synthase complex which is resistant to the antibiotic, oligomycin C. Although, this gene has been used to transform a number of non-pathogenic fungi (42–44), it has yet to be employed with a plant pathogen. A drawback of this system is the need to isolate an *oliC* gene from a resistant mutant of the fungal species to be transformed. Furthermore, genetic characterization of the mutant used as the source of the resistance gene is essential, in order to differentiate between nuclear

Table 2. Plasmids containing hygromycin B phosphotransferase ($hygB^R$) gene constructions

Plasmid	Regulatory sequences	Reference
pAn7-1	*gpd*A promoter and *trp*C terminator sequences from *A. nidulans*	(37)
pDH25	*trp*C promoter sequences from *A. nidulans*	(32)
pITH	promoter isolated at random from *C. heterostrophus*	(9)
pHL1	*hygB*^{R} gene fused to *hsp*70 gene of *U. maydis*	(30)
pCM54	derivative of pHL1 containing an *U. maydis* autonomous replicating sequence	(39)

and mitochondrial mutations. Many fungi are sensitive to oligomycin C and complete inhibition of growth can be expected at $c.$ 0.1–5 μg/ml.

2.1.2 Complementation of nutritional mutants

Few plant pathogens have been transformed by complementation of nutritional mutants. In many cases, auxotrophic mutants are unavailable and the isolation and characterization of suitable mutants can be a laborious procedure. In addition, the absence of a sexual crossing system for many plant pathogens prevents the isolation of recombinants either for genetic analysis or the construction of recipient of strains for the transforming DNA. Some auxotrophic mutants can be isolated relatively easily. By selecting for resistance to toxic substrate analogues of certain enzymes. For example, resistance to the metabolic poison, fluoro-orotic acid, has been used to select auxotrophic mutants (*Pyr* $^-$ mutants) that require uridine or uracil for growth due to a dysfunction in either orotidine-5'-phosphate decarboxylase or dihydroorotase. Complementation of *Pyr* $^-$ mutants by a functional gene restores protrophy, and, thus, provides the basis for selection. Similarly, mutants unable to utilize acetate as a carbon source (*Fac* or *Acu* $^-$ mutants), due to a mutation in the enzyme acetyl-CoA synthese (35), can be isolated by selection for resistance to fluoroacetic acid. Resistance to chlorate can also be exploited to obtain nitrate reductase deficient (*NiaD* $^-$) mutants and a number of plant pathogen mutants have now been transformed using the *niaD* $^+$ gene from *A. nidulans* (17, 22). Procedures for isolating some nutritional deficient mutants are given in *Protocol 1*.

Protocol 1. Procedures for the positive selection of nutritional mutants

1. Mutants may be recovered either spontaneously or after mutagenesis.[a] Sporulating cultures or spores (10^7/ml in saline) can be mutated by irradiating with UV light (254 nm) until 10–30% of the spores remain viable.

2. Plate freshly harvested or irradiated spores directly, or after filtration enrichment (45), on to selection medium. Spread about 10^5/μl spores in 200 ml on each plate.

3. Prepare minimal medium agar[b] to select the mutants by adding the following supplements.

- For Pyr$^-$ mutants. Uridine, 2.44 g/litre and 5-fluoroorotic acid[c], 1 g/litre.

- For *Fac* or *Acu*$^-$ mutants. Fluoroacetic acid[c], 5 g/litre and glucose, 0.1 g/litre to replace glucose alone as the carbon source.

- For *NiaD* mutants. Sodium chlorate, 50 g/litre and sodium glutamate, 1.87 g/litre to replace NO$_3^-$ as the sole source of nitrogen.

4. Purify resistant strains on fresh selection medium.

Protocol 1. *Continued*

5. Test mutants that continue to grow by their requirement for, or inability to utilize specific supplements.

   - *Pyr⁻* mutants require uracil or uridine (add to 10 mM).

   - *Fac* or *Acu⁻* mutants are unable to utilize acetate (5 g/litre) as a carbon source.

   - *NiaD* mutants utilize NO_2^-, NH_4^+, hypoxanthine, and glutamate, (each at 10 mM), but not NO_3^-.

6. Test mutants for reversion to protrophy. Only mutants with a reversion frequency of less than 1 in 10^7 viable spores are suitable to act as recipients for transforming DNA.[d]

[a] Although UV mutagenesis is a safe and convenient method of enhancing mutation rates it is less effective than other mutagens, such as *N*-methyl-*N'*-nitro-*N*-nitrosoguanidine.
[b] Minimal medium requirements vary considerably for different fungi but is usually $NaNO_3$, 6 g; KCl, 0.52 g; KH_2PO_4, KH_2PO_4, 1.52 g; trace elements solution 1 ml; agar 15 g; distilled water to 1 litre. Adjust pH to 6.5 with KOH. After autoclaving add 25 ml 40% (w/v) sterile glucose and 2.5 ml 20% (w/v) sterile $MgSO_4$ per litre. Trace element solution is $ZnSO_4$. $7H_2O$, 1.0 g; $FeSO_4$. $7H_2O$, 8.8 g; $CuSO_4$. $5H_2O$, 0.4 g; $MnSO_4$, 0.15 g; $Na_2B_4O_6$. $10H_2O$, 0.1 g: $(NH_4)6Mo_7O_{24}$, distilled water to 1 litre.
[c] 5-fluoroorotic acid and fluoroacetic acid are extremely toxic and must be handled with extreme care.
[d] It is still not possible to be absolutely sure that a mutation has occurred in the specific gene of interest and not in a regulatory gene. By assaying enzyme activities it is possible to further characterize the mutations, however, this can be a laborious procedure. An easier and quicker approach is to select a number of putative mutants and transform them directly with the chosen selectable gene.

A further nutritional selection utilizes the *amdS⁺* gene of *A. nidulans* (4, 46) which encodes the enzyme acetamidase. This enzyme converts acetamide to acetate and ammonia. Selection of transformants is based on utilization of the generated ammonia as a source of nitrogen and should prove useful with many fungi that are unable to metabolize acetamide. Background growth of untransformed protoplasts, often encountered when using this gene, can be reduced by adding 12.5–15 mM caesium chloride to the medium. However, it should be noted that potassium chloride should *not* be used as the osmotic stabilizer (4).

2.2 Improvement of transformation frequency by additional DNA sequences

DNA transformation of filamentous fungi invariably occurs by either homologous or non-homologous integration of the vector sequences into the chromosomal DNA of the recipient cell. This mode of transformation probably accounts for the relatively low transformation frequencies obtained with filamentous fungi, compared with yeasts and bacteria, where vectors can replicate extrachromosomally. Attempts have been made to enhance transformation frequencies by including additional DNA sequences in the transformation vectors. Usually homologous DNA fragments are used in an attempt to either enhance integration frequencies or confer autonomous replication. On the whole, neither approach has been very successful although a DNA fragment from *A. nidulans*, the *ans*1 sequence, significantly increased transformation frequencies with certain selectable markers. The *ans*1 sequence would appear to enhance transformation by influencing integration events rather than by conferring autonomous replication (47). More recently, an autonomously replicating sequence, UARS1, has been isolated from *U. maydis* (31) and inclusion of part of this sequence in the plasmid pHL1 greatly enhanced transformation frequencies (31, 39).

3. Transformation techniques

Almost all protocols for transforming fungi involve inducing protoplasts to take up DNA. There are three distinct stages to this process.

- Protoplast preparation
- Uptake of DNA
- Regeneration and selection of transformants.

Optimization of transformation conditions should involve critical assessment of each stage as discussed below.

3.1 Protoplast preparation

Isolation of pure, intact protoplasts in sufficient quantities ($> 10^7$) is crucial to the success of a transformation. Young cells, such as freshly harvested spores, germinating spores, germ tubes, or young hyphae, are used as the starting material. The walls of these cells are more susceptible to degradation by lytic enzymes than those of older mature hyphae and the protoplasts generated are easier to release and separate from the mycelial and cell debris. The fungal material used to generate protoplasts is often grown in liquid culture, although as an alternative, germlings or young hyphae can be produced on cellophane discs laid on the surface of agar plates.

During their preparation and isolation, protoplasts need to be stabilized with an osmoticum. Sorbitol is often used, but other alternatives include mannitol, potassium chloride, and magnesium sulphate. All are used at concentrations ranging from 0.6 M to 1.2 M. In some cases, the protoplast preparation can be stored at $-70°C$ for several months without loss of competence (14, 48).

Many different enzyme preparations have been used to generate fungal protoplasts (49). However, the commercial preparation, NovoZyme 234 (Novo BioLabs, Denmark) from *Trichoderma harzianum*, is particularly useful, either alone or in combination with other enzymes. Concentrations between 5 and 25 mg/ml have been used to generate protoplasts from a wide range of fungi and periods of incubation with the enzyme solution vary from 5 minutes to 3 hours or more. The optimum pH for NovoZyme 234 is between 5 and 6. Other enzymes may act synergistically with NovoZyme 234. For example, the addition of cellulase 'onozuka' R-10 (Yakult Honsha Co., Ltd) enhances wall digestion and the release of protoplasts from *G. cingulata f. sp. phaseoli* hyphae (unpublished observations). Other enzymes which have been used to generate fungal protoplasts include Helicase (Industrie Biologique Française), Driselase (Sigma), β-glucuronidase type H2 (Sigma), and chitinase (Sigma). If the cell walls are particularly recalcitrant to digestion, then a pretreatment with thiol reagents (e.g. 25 mM 2-mercaptoethanol or dithiothreitol) and/or EDTA (5 mM) may aid cell wall degradation.

Batches of the same enzyme, particularly NovoZyme 234, vary in their ability to generate transformable protoplasts, although lack of transformability may also result from prolonged exposure to lytic enzymes. It is thus advisable to keep the incubation period with the enzyme to a minimum.

A common mistake when isolating protoplasts from fungi is to use too much starting material resulting in poor recovery of protoplasts contaminated with large amounts of debris. Hyphal debris and undigested cells can usually be removed by filtration through a nylon mesh (e.g. 40–100 μm mesh), glass beads (450–500 μm), or a glass sinter.

Following digestion and removal of debris, the protoplasts are thoroughly washed by centrifugation to remove all traces of the enzyme solution. Many commercial enzyme preparations are crude extracts of fungal culture filtrates and are likely to contain significant contamination with other enzymes, such as DNAases, which affect transformation efficiency. Care should also be taken to avoid lysis of the protoplasts during the centrifugation step or whilst resuspending the protoplasts. Centrifugation at 100–1000 g for 10 min is generally adequate to pellet most fungal protoplasts and three to four washes should be sufficient to provide a clean preparation. If washing the protoplasts is a problem then they may be recovered by flotation (4).

Finally, it should be emphasized that preparing fungal protoplasts for transformation is not difficult but requires care and constant monitoring. A light microscope equipped with phase contrast optics is essential for assessing the quality of the protoplast preparation.

3.2 DNA uptake

Protoplasts are induced to take up DNA molecules by treatment with polyethylene glycol (PEG) and Ca^{2+} ions. There are numerous modifications to the basic methods and at least two or three different protocols should be tried when developing a transformation system. Unfortunately, it is not possible to generalize about the optimum conditions, even within the same species. However, there are three methods, basically differing in the size and concentration of PEG and the concentration of Ca^{2+} ions (*Protocol 2*), that have been used either directly, or after minor modifications, with a wide range of fungi.

Protocol 2. Protoplast transformation

Procedure A (adapted from ref. 47)

1. To 100 μl of protoplasts (c. 10^7) in 0.6 M KCl, 50 mM $CaCl_2$, add 5 μl of DNA (1–20 μg) dissolved in TE buffer,[a] 25 μl 25% PEG 6000 in 50 mM $CaCl_2$, 10 mM Tris–HCl pH 7.5 (filter sterilized). Leave on ice for 20 min.

2. Add 1 ml of the same PEG solution and mix. Incubate with gentle agitation for 5 min at room temperature.

3. Add 2 ml 0.6 M KCl, 50 mM $CaCl_2$, and mix.

4. Either
   - add aliquots of the transformation mixture to 5 ml molten regeneration agar[b] at 48°C, and pour on to bottom regeneration medium, or
   - recover the protoplasts by centrifugation, resuspend in a suitable osmoticum, and plate directly on to regeneration medium.

Procedure B (adapted from ref. 48)

1. To 100 μl protoplasts (c. 10^7) in STC1 buffer[c] containing 1.2% DMSO add 5 μl (1–20 μg) DNA in TE buffer containing 5 μg heparin.

2. Leave on ice for 30 min.

3. Slowly add 1 ml filter sterilized 40% PEG 4000 in STC1 buffer, mix, and leave at room temperature for 20 min.

4. Plate out the transformation mixture as described in Procedure A, step 4.

Procedure C (adapted from ref. 50)

1. To 100 μl protoplasts (c. 10^7) in STC1 buffer add up to 25 μl (1–20 μg) DNA in STC1 buffer.

2. Leave at room temperature for 25 min.

3. Add 1.25 ml filter sterilized 60% PEG 4000 in STC1 buffer in succeessive 200, 200, and 850 μl aliquots and incubate at room temperature for a further 20 min.

4. Plate out the transformation mixture as described in Procedure A, step 4.

[a] TE buffer is 10 mM Tris–HCl, pH 7.5, 1 mM EDTA.
[b] Regeneration agar contains an appropriate osmotic stabilizer (e.g. 0.6 M KCl or 1 M sorbitol).
[c] STC1 buffer is 1 M sorbitol, 50 mM Tris–HCl, 50 mM $CaCl_2$, pH 7.5.

One to twenty μg of DNA is added to the protoplast preparation (c. 10^7 protoplasts/transformation). The DNA may be suspended in heparin and, in some cases, spermidine is also added (49). Other adjuvants which may influence the uptake of DNA and increase transformation frequencies include DMSO (48) and 2-mercaptoethanol (31). The protoplasts are incubated with the DNA, either at room temperature or on ice, for 10–30 min before the PEG solution is added.

DNA prepared using CsCl gradients usually gives higher transformation frequencies than that obtained from plasmid mini-preps.

Treatment of protoplasts with PEG and Ca^{2+} ions causes them to aggregate, fuse, and to become temporarily permeable to DNA molecules. PEG (mol. wt 3350 to 6000) is used at concentrations between 25 and 66% (w/v) and $CaCl_2$ is present at concentrations ranging from 10 mM to 100 mM. The size, concentration, and volume of PEG, and the period of exposure to the PEG solution, vary in different transformation protocols and for different fungi. It is advisable to add the PEG solution slowly in increasing aliquot volumes and to mix gently before adding the next aliquot. PEG can be toxic to some fungi and the length of exposure of protoplasts to the transformation solution should be kept to a minimum (< 30 min).

Electroporation (51) offers an alternative method for transferring DNA into protoplasts but has not been used extensively with filamentous fungi.

3.3 Regeneration and selection of transformants

Following transformation, protoplasts are regenerated and exposed to the appropriate selection pressure. The efficiency of regeneration can vary enormously; between 0.1 and 70% regeneration can be expected. Failure to regenerate is likely to be due to excessive physical damage or stress during protoplast isolation, the lethal effects of the PEG treatment, or the lack of nuclei in some protoplasts. Protoplasts are plated either in an agar overlay or directly on to a selective medium containing an appropriate osmotic stabilizer; both methods should be tried. The concentration of agar in the overlay may also affect the number of transformants recovered. For example, with *A. nidulans*, increasing the overlay agar concentration from 0.9% to 2% enhances the recovery of transformants by 3–4 fold (47). Agarose can be used instead of agar in the overlay medium.

Where selection of transformants is based on the expression of a drug resistance gene, a period of recovery is often required before the protoplast are exposed to the drug (42). This allows the gene to be expressed at the level required to confer a resistant phenotype. The duration of the recovery period varies with the fungus and with the drug. For the *hygB*^{R} gene this is between 3 and 24 h. Extending the length of the recovery period can lead to a reduction in the number of transformants and may also allow non-transformed hyphae to escape inhibition and outgrow the transformed protoplasts.

The simplest way to allow protoplasts to recover before exposure to the inhibitor is to plate on to a non-selective medium for an appropriate period of time, before overlaying with medium containing a suitable concentration of the drug. Alternatively, protoplasts can be incubated in liquid (or ultra-low melting point agarose containing) media before being plated on agar media containing the drug. In one procedure for selecting *hygB*^{R} transformants of *U. maydis* a selection system was used where the drug was incorporated into a lower layer of agar which, once set, was overlayed with unamended medium. After a period of recovery the transformation mixture was plated directly on to the freshly prepared selection plate (30). When selecting for drug resistance, it is essential to pour agar plates and overlays on a level surface. If a solution of the drug is to be added to regenerating protoplasts then the whole plate should be evenly covered by holding the plate at an angle and rotating slowly until the entire surface is covered. Uneven distribution of the drug results in transformants being killed in regions where excessive amounts of the drug are present and to growth of untransformed protoplasts where there is insufficient inhibitor.

Two complete transformation methods are provided in *Protocols 3* and *4*. *Protocol 3*, used for *U. maydis*, was adapted from two published methods (30, 31) and uses the plasmid pCM54 carrying the *hygB*[R] gene (see *Table 2*).

Protocol 3. Transformation of *U. maydis* using pCM54

1. Grow an overnight shaken culture to 10^7 cells/ml in 100 ml YEPD[a] medium at 32°C.

2. Centrifuge at 2000 g for 10 min.

3. Wash with 12 ml SCS buffer[b] and resuspend cells in 1 ml protoplasting solution.[c] Incubate for 5–10 min with gentle shaking, make sure the cells are not clumped and check wall digestion under the microscope.

4. Centrifuge at 1000 g for 10 min just before all the cells are completely converted to protoplasts.

5. Wash the protoplasts twice with SCS buffer (by this time all cell wall debris and lysed protoplasts should have been removed).

6. Wash protoplasts once and resuspend in 1 ml STC2 buffer.[d] Store protoplasts on ice until required.

7. To transform, mix 1 μl of DNA in TE buffer[e] with 3 ml heparin solution[f] in a 2 ml tube. Add 50 μl (*c.* 10^6) of protoplasts, gently mix, and leave on ice for 10 min.

8. Gradually add the PEG solution[g] stepwise in 50 μl, 150 μl, and 300 μl aliquots. Use a disposable pipette tip with the end cut off to dispense accurate quantities of the viscous PEG solution. Protoplasts are completely mixed with the PEG solution before adding the next aliquot. Incubate for a total period of 30 min.

9. Add transformation mixture to 5 ml molten regeneration medium[h] (48°C) and overlay on to plates containing 15 ml regeneration agar. Ensure plates are poured on a level surface and that volumes are accurate.

10. Incubate plates, top surface uppermost, for 4 h at 32°C then add 500 μl hygromycin solution,[i] making sure that the whole plate is evenly covered by the hygromycin solution.

Protocol 3. *Continued*

11. Incubate plates face-up on a level surface, at 32°C. After 24 h seal plates with parafilm, invert, and continue to incubate.[j]

---

[a] YEPD medium is 1% yeast extract, 2% Bacto-peptone (Difco), 2% dextrose (or sucrose). Add 2% agar (Difco Bacto-agar) for solid medium.
[b] SCS buffer is 1 M sorbitol, 20 mM sodium citrate (pH 5.8).
[c] Protoplasting solution is 25 mg NovoZyme 234/ml in SCS buffer, filter sterilized and stored at −20°C.
[d] STC2 buffer is 1 M sorbitol, 10 mM Tris–HCl, 100 mM CaCl$_2$ (pH 7.5).
[e] See footnote [a] to *Protocol 2* for TE buffer.
[f] Heparin solution is 5 mg heparin ml in STC, filter sterilized, and stored at −20°C.
[g] PEG solution is 66% (w/v) PEG 3350 or 4000 in 25 mM CaCl$_2$, 25 mM Tris–HCl pH 7.5, filtered through 1.2 μm filter, and stored at −20°C.
[h] Regeneration medium is YEPD medium containing 1 M sorbitol and 2% agar. The overlay can contain 0.35% agarose instead of agar (this aids the collection of colonies).
[i] Hygromycin B solution is 8 mg hygromycin B/ml in 1 M sorbitol, filter sterilized and stored at −20°C.
[j] Transformants are visible microscopically after *c.* 30 h and colonies can be picked off 2–3 days later. Note that colonies submerged in overlay grow slower than those on the surface.

Protocol 4, based on methods developed for *A. nidulans* (3), is used for *G. cingulata f. sp. phaseoli*, and utilizes another *hygB*[R] gene containing vector, pAN7–1. Typical selection plates from these two transformation protocols are illustrated in *Figure 1*.

Protocol 4. Transformation of *G. cingulata f. sp. phaseoli* using pAn7-1

1. Cut circles of cellophane using the base of a Petri dish as a template. Sterilize by autoclaving in distilled water and transfer cellophane circles to plates containing ANM[a] agar.

2. Spread *c.* 500 μl of a conidial suspension (10^6 conidia/ml) on to the surface of the cellophane discs and incubate for 2 days at 25°C. Make sure that the mycelium is not over grown; a thin covering of mycelium on the cellophanes is best. Six cellophane circles should give enough protoplasts for at least 4 transformations.

3. Transfer discs to a clean Petri dish containing 30 ml SCS buffer[b] and incubate for 30 min. During this period scrape the mycelium off the cellophane circles with a sterile spatula. Squeeze against the edge of the Petri dish and transfer to 20 ml protoplasting solution[c] and incubate at room temperature. Monitor the release of protoplasts and proceed to the next step when the majority of cells have lost their walls (usually after *c.* 50–60 min).

4. Filter the protoplasts through a 250 μm nylon mesh and through a bed of acid-washed glass beads (Sigma type V, 450–500 μm) supported on a 100 μm nylon mesh. Agitate surface of the beads with a sterile spatula and wash the protoplasts through the bed with *c.* 60 ml STC buffer.[d]

Protocol 4. *Continued*

5. Centrifuge the protoplast preparation at 1500 *g* for 10 min and wash twice in STC3 buffer. During washing gently resuspend protoplasts ensuring that clumps of protoplasts are dispersed.

6. Resuspend protoplasts in 800 μl STC3 buffer and keep on ice.

7. Add pAN7–1 (up to 10 μl containing 1–20 μg/ml in TE buffer) to 100 ml protoplast preparation (*c.* 10^7 protoplasts/ml) and leave on ice for 30 min.

8. Add 25 μl PEG solution[f] mix well, and leave on ice for 30 min. Add 1 ml of the PEG solution, mix, and incubate at room temperature with gentle shaking for 15 min.

9. Add the transformation mixture to 50 ml 0.1 × ANM liquid medium containing 2% ultra-low melting point agarose (Sigma, type IX) and 1 M sorbitol. Incubate on a rotary shaker (150 r.p.m.) at 25°C for 18–24 h.

10. Overlay 5 ml of protoplasts in agarose on to 20 ml regeneration medium containing 62.5 μg hygromycin B/ml. Cool the plates at 4°C for 30 min to solidify the agarose, making sure that plates are kept level.

11. Incubate at 25°C. Microscopic transformants will appear after 3 or 4 days and colonies appear after *c.* 6 days.

[a] ANM medium is 2% malt extract, 0.1% Bacto-peptone, and 2% glucose. Add 2% agar for solid medium.
[b] See footnote [b] to *Protocol 3* for SCS buffer.
[c] Protoplasting solution is SCS buffer containing 7 mg NovoZyme 234 and 7 mg cellulase R-10/ml.
[d] STC3 buffer is 1 M sorbitol, 10 mM $CaCl_2$ and 10 mM Tris pH 7.5.
[e] See footnote [a] to *Protocol 2* for TE buffer.
[f] PEG solution is 25% (w/v) PEG 6000 in 10 mM $CaCl_2$ and 10 mM Tris pH 7.5, filtered through a 1.2 μm filter and stored at −20°C.
[g] Regeneration medium is ANM medium containing 1 M sorbitol and 2% agar.

3.4 Whole cell techniques

There are alternative methods to make fungi competent to take up DNA that avoid the use of protoplasts although these usually result in the recovery of fewer transformants. In one method, the alkali metal ion lithium is employed along with PEG, while another method used physical damage.

Transformation of intact cells treated with Li^+ ions was first demonstrated for *S. cerevisiae* (52) and subsequently for *N. crassa* (53). Modifications of this method have been adopted for the agaric fungus, *Coprinus cinereus* (54) and the plant pathogens, *Colletotrichum capsici* (10), *C. trifilii* (12), *Fusarium solani f. sp. pisi* (10), and *U. violacea* (36). In all these cases, germinating spores or mycelium were the recipients for the DNA and lithium acetate was used in conjunction with PEG.

The second method which, so far, has only been employed with the yeast, *S. cerevisiae*, is highly attractive because of its simplicity (55). Permeability

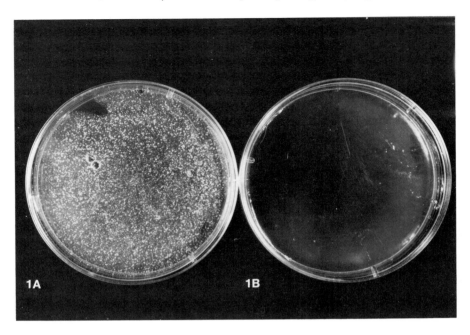

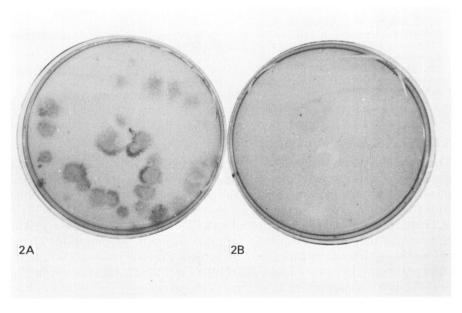

Figure 1. Selection for *hygB*[R] transformants of either *U. maydis* (1) or *G. cingulata f. sp. phaseoli* (2). In 1A and 2A, protoplasts received 500 ng pCM54 and 1 μg pAn7-1 respectively. 1B and 2B protoplasts received no DNA. Transformants were as described in Tables 5 and 6.

changes, which allow DNA to be taken up into the cell, are induced by vortexing the cells with glass beads in the presence of the transforming DNA. Although transformation frequencies are low, this is more convenient than conventional methods and could be applied to many fungi from which protoplasts are difficult to obtain. In addition to these methods, recent developments in the use of microprojectile bombardment to transform yeast (56) are worth investigating, particularly with biotrophic pathogens which cannot be cultured *in vitro*.

4. Strain storage

Most fungal transformants are stable because the vector sequences integrate into the host genome. Nevertheless, it is essential that both recipient strains and transformants are stored to avoid working with strains that have altered during subculturing (e.g. loss of pathogenicity) or with transformants in which the plasmids have rearranged or have been deleted.

There are many methods for the long term storage of fungi and three relatively simple examples are given in *Protocol 5*. In addition, some fungi may be snap frozen in liquid nitrogen and stored at $-70°C$ in media containing 25% (v/v) glycerol.

Protocol 5. Storage of strains and transformants

A. *Silica gel method* (adapted from ref. 57)

1. Use glass universal bottles or similar with air-tight caps (rubber seals replaced with foil). Fill bottles a third full with silica gel and sterilize for 90 min, at 180°C. Store in a desiccator until required.

2. Prepare a 5% solution of skimmed milk powder in distilled water, dispense in 2 ml aliquots, and autoclave for 10 min at 15 p.s.i. Store at 4°C.

3. Prepare a dense suspension of a fresh culture (mycelium or spores) in skimmed milk and cool silica gel containers on ice for at least 30 min.

4. Add suspension to the silica gel until *c*. three-quarters is wetted then return bottle to ice for at least 15 min.

5. Leave at room temperature until crystals are readily separated. Screw cap down firmly and store at room temperature or 4°C. A small sample of crystals can be taken as required.

B. *Filter paper method* (adapted from ref. 58)

This method is simple and allows storage of large numbers of strains in an organized filing system without taking up too much space.

1. Place sterile filter paper discs (13 mm diameter) on to a freshly inoculated agar plate and allow fungus to grow into the paper.

2. Peel papers from the plate and air dry in a Petri dish for 2–3 days.

Protocol 5. *Continued*

3. The paper discs can be stored in sterile stamp envelopes and filed in waterproof boxes. The discs are stored at either 4°C or −20°C.

C. *Water immersion method* (adapted from ref. 57)

This is probably the simplest way to store cultures and many fungi can be recovered after more than a year.

1. Grow a fresh culture of the fungus on slants in universal bottles.

2. Pour sterile distilled water over culture until it is completely submersed.

3. Replace cap and ensure that an air-tight seal is made. Store bottles at room temperature.

5. Cloning pathogenicity genes by complementation

Strategies for cloning particular genes of interest, such as pathogenicity-related genes, can be considered once a transformation system has been developed. The vector should carry at least one unique restriction enzyme site for insertion of DNA fragments. If transformation frequencies are low, then vector improvement by, for example including autonomously replicating sequences or *ans*1 type sequences could be considered. A more common approach to overcome this problem, however, is to construct a cosmid derivative of the vector, which enable larger DNA fragments to be cloned. This significantly reduces both the number of recombinants required to represent the whole genome of the fungus and the number of transformants which need to be screened. An additional advantage of using cosmids is that clones can be rescued relatively easily by either a sib-selection procedure (48) or by packaging DNA from the transformant into the lambda-phage particles (50). However, it should be noted that cosmid gene banks can be unstable and some clones may undergo gross rearrangements. Furthermore, cosmid bank DNA is difficult to propagate in large amounts in *E. coli* and reduced transformation frequencies can be expected. Nevertheless, cosmids are still the most practical way to clone fungal genes by expression. To date, only two pathogenicity-related genes have been isolated by expression; a mating type allele from *U. maydis*, which is involved in determining the pathogenic habit (60) and a gene from *Nectria haematococca* which encodes pisatin demethylase (59), an enzyme that detoxifies the pea phytoalexin, pisatin. Both genes were isolated from cosmid gene banks. The *U. maydis* mating type allele was identified by expression in a homologous diploid host, and the pisatin demethylase gene was isolated by heterologous expression in the non-pathogenic fungus, *A. nidulans*. In both cases, the initial selection did not require screening of transformants on plants, and mutants were not required as recipients of the transforming DNA.

John Hargreaves and Geoffrey Turner

Acknowledgements

We are extremely grateful to Professor Bill Holloman for providing pCM54 and to Professor Cees van den Hondel for providing pAn7-1. We also wish to thank Professor Sally Leong for the plasmid pHL1 which was used to develop transformation of *Ustilago* in our laboratory.

References

1. Fincham, J. R. S. (1989). *Microbiol. Rev.*, **53**, 148.
2. Case, M. E., Schweizer, M., Kushner, S. R., and Giles, N. H. (1979). *Proc. Natl. Acad. Sci. USA*, **76**, 5259.
3. Ballance, D. J., Buxton, F. P., and Turner, G. (1983). *Biochem. Biophys. Res. Commun.*, **112**, 284.
4. Tilburn, J., Scazzocchio, C., Taylor, G. G., Zabicky-Zissman, J. H., Lockington, R. A., and Wayne-Davies, R. (1983). *Gene*, **26**, 205.
5. Huang, D., Bhairi, S., and Staples, R. C. (1989). *Curr. Genet.*, **15**, 411.
6. Upchurch, R. G., Young, L. A., and Walker, D. C. (1989). 15th Fungal Genetics Conference, Pacific Grove, California, Abstr. 31.
7. van Engelenberg, F., Smit, R., Goosen, T., van den Broek, H., and Tudzynski, P. (1989). *Appl. Microbiol. Biotechnol.*, **30**, 364.
8. Turgeon, B. G., Garber, R. C., and Yoder, O. C. (1985). *Molec. Gen. Genet.*, **201**, 450.
9. Turgeon, B. G., Garber, R. C., and Yoder, O. C. (1987). *Mol. Cell. Biol.*, **7**, 3297.
10. Soliday, C. L., Dickman, M. B., and Kolattukudy, P. E. (1989). *J. Bacteriol.*, **171**, 1942.
11. Pannaccione, D. G., McKiernan, M., and Hanau, R. M. (1988). *Molec. Plat–Microbe Interactions*, **1**, 113.
12. Dickman, M. B. (1988). *Curr. Genet.*, **14**, 241.
13. Churchill, A. C. L. and VanAlfen, N. K. (1989). 15th Fungal Genetics Conference, Pacific Grove, California, Abstr. 30.
14. Oliver, R. P., Roberts, J. N., Harling, R., Kenyon, L., Punt, P. J., Dingemanse, M. A., et al. (1987). *Curr. Genet.*, **12**, 13.
15. Leslie, J. F. and Dickman, M. B. (1989). 15th Fungal Genetics Conference, Pacific Grove, California, Abstr., 9.
16. Kistler, H. C. and Benny, U. K. (1988). *Curr. Genet.*, **13**, 145.
17. Malardier, L., Daboussi, M. J., Julien, J., Roussel, F., Scazzocchio, C., and Brygoo, Y. (1989). *Gene*, **78**, 147.
18. Salch, Y. P. and Beremand, M. N. (1989). 15th Fungal Genetics Conference, Pacific Grove, California, Abstr., 14.
19. Marek, E. T., Schardl, C. L., and Smith, D. A. (1989). *Curr. Genet.*, **15**, 421.
20. Rodriguez, R. J. and Yoder, O. C. (1987). *Gene*, **54**, 73.
21. Daboussi, M. J., Djeballi, A., Gerlinger, C., Blaiseau, P. L., Bouvier, I., Cassan, M., et al. (1989). *Curr. Genet.*, **15**, 453.
22. Henson, J. M., Blake, N. K., and Pilgeram, A. L. (1988). *Curr. Genet.*, **14**, 113.

23. Farman, M. L. and Oliver, R. P. (1988). *Curr. Genet.*, **13**, 327.
24. Cooley, R. N. and Caten, C. E. (1989). EMBO Workshop: Molecular Biology of filamentous fungi, Espoo, Finland, Abstr. 17.
25. Parsons, K. A., Chumley, F. G., and Valent, B. (1987). *Proc. Natl. Acad. Sci. USA*, **84**, 4161.
26. Leong, S., Skinner, D., Leung, H., Karjalainen, R., and Ellingboe, A. (1989). 15th Fungal Genetics Conference, Pacific Grove, California, Abstr. 13.
27. Royer, J. C., Hubbes, M., and Horgen, P. A. (1989). 15th Fungal Genetics Conference, Pacific Grove, California, Abstr. 33.
28. Blakemore, E. J. A., Dobson, M. J., Hocart, M. J., Lucas, J. A. and Peberdy, J. F. (1989). *Curr. Genet.*, **16**, 177.
29. Cooley, R. N., Shaw, R. K., Franklin, F. C. H., and Caten, C. E. (1988). *Curr. Genet.*, **13**, 383.
30. Wang, J., Holden, D. W., and Leong, S. A. (1988). *Proc. Natl. Acad. Sci. USA*, **85**, 865.
31. Tsukuda, T., Carleton, S., Fotheringham, S., and Holloman, W. K. (1988). *Mol. Cell. Biol.*, **8**, 3703.
32. Cullen, D., Leong, S. A., Wilson, L. J., and Henner, D. J. (1987). *Gene*, **57**, 21.
33. Banks, G. R. (1983). *Curr. Genet.*, **7**, 73.
34. Banks, G. R. and Taylor, S. Y. (1988). *Mol. Cell. Biol.*, **8**, 5417.
35. Hargreaves, J. A. and Turner, G. (1989). *J. Gen. Microbiol.*, **135**, 2675.
36. Bej, A. K. and Perlin, M. H. (1989). *Gene*, **80**, 171.
37. Punt, P. J., Oliver, R. P., Dingemanse, M. A., Pouwels, P. H., and van den Hondel, C. A. M. J. J. (1987). *Gene*, **56**, 117.
38. Gritz, L. and Davies, J. (1983). *Gene*, **25**, 179.
39. Fotheringham, S. and Holloman, W. K. (1989). *Mol. Cell Biol.*, **9**, 4052.
40. Orbach, M. J., Porro, E. B., and Yanofsky, C. (1986). *Mol. Cell Biol.*, **6**, 2452.
41. Judelson, H. S. and Michelmore, R. W. (1989). *Gene*, **79**, 207.
42. Ward, M., Wilkinson, B., and Turner, G. (1986). *Molec. Gen. Genet.* **202**, 265.
43. Ward, M., Wilson, L. J., Carmona, C. L. and Turner, G. (1988). *Curr. Genet.*, **14**, 37.
44. Bull, J. H., Smith, D. J., and Turner, G. (1988). *Curr. Genet.* **13**, 377.
45. van Hartingsveldt, W., Mattern, I. E., van Zeijl, C. M. J., Pouwels, P. H., and van den Hondel, C. A. M. J. J. (1987). *Molec. Gen. Genet.*, **206**, 71.
46. Hynes, M. J., Corrick, C. M., and King, J. A. (1983). *Mol. Cell Biol.*, **3**, 1430.
47. Ballance, D. J. and Turner, G. (1985). *Gene*, **36**, 321.
48. Vollmer, S. J. and Yanofsky, C. (1986). *Proc. Natl. Acad. Sci. USA*, **83**, 4869.
49. Peberdy, J. F. (1985). In *Fungal protoplasts, applications in biochemistry and genetics* (ed. J. F. Peberdy and L. Ferenczy), p. 31. Marcel Dekker, New York.
50. Yelton, M. M., Hamer, J. E., and Timberlake, W. E. (1984). *Proc. Natl. Acad. Sci. USA*, **81**, 1470.
51. Richey, M. G., Marek, E. T., Schardl, C. L., and Smith, D. A. (1989). *Phytopathology*, **79**, 844.
52. Ito, H., Fukuda, Y., Murata, K., and Kimura, A. (1983). *J. Bacteriol.*, **153**, 163.
53. Dhawale, S. S., Paietta, J. V., and Marzluf, G. A. (1984). *Curr. Genet.*, **8**, 77.
54. Binninger, D. M., Skrzynia, C. S., Pukkila, P. J., and Casselton, L. A. (1987). *EMBO J.*, **6**, 835.
55. Costanazo, M. C. and Fox, T. D. (1988). *Genetics*, **120**, 667.
56. Johnston, S. A., Anziano, P. Q., Shark, K., Sanford, J. C., and Butow, R. A. (1988). *Science*, **240**, 1538.

57. Smith, D. and Onions, A. H. S. (1983). *The preservation and maintenance of living fungi.* Commonwealth Mycological Institute, London.
58. Valent, B., Crawford, M. S., Weaver, C. G., and Chumley, F. G. (1986). *Iowa State Journal of Research*, **60**, 569.
59. Weltring, K-M., Turgeon, B. G., Yoder, O. C., and VanEtten, H. D. (1988). *Gene*, **68**, 335.
60. Kronstand, J. W. and Leong, S. A. (1989). *Proc. Natl. Acad. Sci. USA*, **86**, 978.

9

Nematodes

HOWARD J. ATKINSON

1. Introduction

Plant parasitic nematodes are either ecto- or endoparasites of plants with the majority of species attacking roots. They develop from an egg, through four juvenile stages (J1–J4), to a mature adult in a lifecycle lasting from a few weeks to several months. Most species are less than 2 mm in length and use a hollow stylet both to pierce plant cell walls and to withdraw cell contents. Only the stylet is inserted by ectoparasites whereas endoparasites either migrate intracellularly using the stylet to perforate cell walls or they move bodily between cells. Some endoparasites migrate short distances before feeding whereas others move continually or rely on host growth to assist their distribution within the plant.

Nematodes from several genera, including both the economically important cyst and root-knot nematodes, modify plant cells into feeding sites able to support sedentary females. Such individuals grow rapidly and within a few weeks have increased by up to 1000-fold in volume ensuring a high fecundity (1). This brief introduction will focus on cyst and root-knot nematodes but detailed accounts of all the economic nematodes of world agriculture are widely available (2–5).

2. Economic status

Cyst nematodes (principally *Heterodera* and *Globodera* spp.) are key pests of major crops. *Heterodera glycines* is the principal pathogen of soybean in USA with an economic effect that may lie between $500–1000 m/year. *Heterodera shachtii* (Beet cyst nematode) causes major problems to sugar beet growers in the EC and parts of the USA. *Heterodera avenae* (cereal cyst nematode) is a cosmopolitan pathogen of cereals with particular importance in more arid soils, for example parts of Australia. Potato cyst nematodes *Globodera rostochiensis* and *G. pallida* occur in many countries and they can be highly damaging. They are estimated to impose an annual cost of £10–50m to the UK potato industry.

Root-knot nematodes (*Meloidogyne* spp.) are associated with tropical and subtropical soils and few other pathogens surpass them in importance to world agriculture. There are many species but five are responsible for the majority of crop damage. Indeed, *M. incognita* is estimated to account for about 66% of all economic loss caused by root-knot nematodes. Severity of crop damage varies but overall losses of 11–25% have been estimated for a wide range of crops in major geographical regions of the tropics (6).

3. Control

Current control practices rely on crop rotation, chemicals, or resistant cultivars, often used in an integrated manner (7, 8). There is, however, an urgent need to improve control for the following reasons:

(a) Nematicides are among the most unacceptable compounds in widespread use. One carbamate, aldicarb, and its breakdown products are highly toxic to mammals and have polluted groundwater in USA and presumably other areas where this pesticide is widely used.

(b) Cultural control includes hidden losses that are unacceptable to specialist growers or those with few alternative, economic crops.

(c) Resistant cultivars are not always available and may produce lower yields than the best susceptible cultivars, so, again, they involve hidden losses unless economic populations are known to infest the growing area.

The inadequacy of current crop protection methods justifies research towards novel approaches to the control of nematodes.

4. Biology

4.1 Cyst nematodes

The cyst is the tanned body wall of a former female that encloses much of her egg production. It protects the eggs from harsh environmental conditions and many predators while the dispersal of encysted eggs ensures the involvement of enough individuals to found a competent reproductive unit. The annual hatch of juveniles from encysted eggs in the absence of a host typically varies with species from 10–66%/year. This allows the persistent forms, such as *G. pallida*, to survive up to 10–15 years in soil without a host. Some species are stimulated to hatch by specific chemicals exuded by host roots (9, 10). One moult occurs before each J2 worm hatches from its egg and emerges into the soil through natural orifices in the cyst wall. It moves a short distance to a nearby root, invades, and establishes a feeding site within a host plant. Development to female involves feeding of all stages including adult whereas the smaller male is formed after feeding to the end of J3 stage only. The sedentary, saccate female protrudes on to the root surface and is fertilized by the vermiform, active male before her body wall is tanned to form the cyst. One complete and a second, partial generation is completed on the potato crop in Northern Europe with a typical 10–20-fold population increase from an economic threshold of about 10 viable eggs/g soil at planting (3, 7, 8). Life span of the host and soil temperature influence the number of generations, but in warm soils reproduction of species such as *Heterodera glycines* may be limited by high soil temperatures or more often by soil aridity.

Protocol 1 summarizes simple methods for maintaining cultures as an alternative to recovery from severe field infestations. *Protocol 2* describes the recovery of nematodes from soil. It is essential to ensure all acquisitions and colonies conform to relevant plant health regulations.

4.2 Root-knot nematodes

Some species are parthenogenetic which ensures that all individuals have the potential to contribute directly to population increase and avoid the difficulty of finding a mate at low densities or in arid soils. The major economic species have a much wider host range than is typical of a cyst nematode. The infective juvenile hatches and migrates to the root meristem. Three moults occur to produce an adult female. She feeds extensively and produces eggs through a lesion into a protective gelatinous matrix on the root surface. Juveniles may hatch and re-invade the host or survive, often unhatched, within the protective matrix until growth of a subsequent host. A simple method for maintaining root-knot nematodes is given in *Protocol 1 C*.

Protocol 1. Basic methods for maintaining nematode colonies

A. *Soybean cyst nematode (SCN) under glasshouse conditions*

1. Mix sand and loam (50:50; no peat) with infected soil to 20 viable eggs/g or about 20 cysts/100 g soil (see *Protocol 2* for extraction).

2. Germinate 3 soybeans per 9 cm pot at 25–30°C.

3. Place pots on perlite. Water perlite daily and the soil after 10 days.

4. Collect roots and attached soil at 40 days and store at 25–32°C for 28 days and subsequently at 5°C until use. Expected yield is about 1×10^6 eggs/25 pots.

B. *Potato cyst nematode (PCN)*

1. As for SCN except grow potato plants at 15–22°C. If outdoors, use 2 tubers per 2 kg pot and about 50 viable eggs/g soil.

2. Leave soil after harvest for about 3 months, bag soil, and store at 5°C.

C. *Root-knot nematodes (RKN)*

1. Re-pot one four-week-old tomato plant (e.g. cv Moneymaker, Rutgers) adding 10% of a galled root system carrying egg masses. Grow as for SCN.

2. Check at 6–8 weeks for complete embryonation of > 50% of eggs within the

Protocol 1. *Continued*

masses. When ready, extract eggs (see *Protocol 2*) for a yield of about $0.1–0.5 \times 10^6$ eggs/plant.

Protocol 2. Extraction of nematodes

These methods are simple procedures and do not require specialized equipment. More complex methods with higher extraction efficiencies are available (5, 11).

A. *Potato cyst nematodes (PCN)*

1. Mix 1 kg of air-dried, infected soil in a bucket of water, swirl vigorously, stand for 25 sec and pour through a 1 mm sieve on to a 250 μm sieve.

2. Stand for 5 min and collect floating debris and dry. Roll cysts from dry debris using an inclined metal tray and/or by flotation over acetone. Store at 4°C.

3. Place cysts on a 30 μm nylon sieve at 15–20°C for 7 days in water then in potato-root diffusate[a] for 28 days. Collect hatched J2 twice weekly.

B. *Soybean cyst nematode (SCN)*

1. Decant from damp soil as for PCN, step 1.

2. Centrifuge in water for 2 min at about 1000 g, discard supernatant. Re-centrifuge in sucrose (specific gravity 1.22) for 2 min. Transfer floating cysts into a large volume of water and recollect on a 250 μm sieve.

3. Stir cysts in water with a magnetic bar until they are ruptured and eggs released. Centrifuge as in step 2, resuspend pellet in sucrose, and re-centrifuge to collect floating eggs.

4. Recollect eggs on to 30 μm nylon sieve,[b] treat with 0.1% aqueous Malachite Green for 20 min, and rinse several times in water.

5. Hatch on 30 μm sieve at 25–30°C in water or soybean-root diffusate. Typical hatching rates are 5–20%/wk. Use within a few days of hatching.

C. *Root knot nematodes (RKN)*

1. Take the galled roots grown as described in *Protocol 1*. Wash galled roots and cut into about 5 cm pieces.

2. Mix vigorously in 2% sodium hypochlorite for 2 min and immediately disperse through 150 μm sieve to a large volume of water. Collect eggs and debris on 30 μm sieve and collect eggs from other debris using sucrose. Centrifuge as for SCN, step 2.

3. Hatch in water as for SCN. Expect < 50% hatch, collect at 4–7 days, and use within 1–3 days of collection.

Protocol 2. *Continued*

a Collect by percolating sufficient water through potted, young potato plants in soil or perlite. Collect about 50 ml/pot every 2 days, filter, and store in plastic at 4°C.
b Nylon mesh trapped between rings made from 3.1 and 3.8 cm diameter plastic piping used in household plumbing.

5. Pathology

5.1 Effects on whole plant and yield

High densities of cyst nematode cause stunted plants with a small root system. The diseased plants show symptoms of mineral deficiencies in their leaves and wilt readily. Yield losses are related to the severity of parasitism above a tolerance limit and can be substantially greater than 50% for some species. Root-knot nematodes cause many of the effects described for cyst nematodes with the addition that the root system is often heavily galled with increased accessibility to secondary pathogens (7, 11, 12).

5.2 Histopathology

5.2.1 Cyst nematodes

Soybean cyst nematode typically invades in the zone of root elongation and the animal moves intracellularly by cutting plant cell walls with its stylet. It selects an initial feeding cell close to the endodermis and induces a syncytial cell system from which it feeds. *Heterodera schachtii* shows a repeated sequence of three stages.

(a) Stylet insertion into the plant plus salivary secretion.

(b) Pharyngeal pumping during ingestion.

(c) Stylet withdrawal from the plant cell and reaccumulation of secretions close to the rear of the stylet.

The secretions are partly responsible for the formation of a feeding tube which extends into the syncytium (13–15). The susceptible plant often shows a necrotic response during intracellular migration which is normally in the region of root elongation. The parasite becomes established after selecting an initial feeding cell in an endodermal or an inner cortical cell. Cell walls of adjacent cells show wall breakdown with a reduced number of plasmodesmata between the syncytium and unmodified cells and some necrosis around its perimeter. The syncytium forms a wedge that progresses into the central vascular tissue and finger-like protuberances occur in close proximity to the xylem, and to a lesser extent, phloem tissue (16). The syncytium establishes itself as a transfer cell system at about 7 days post-invasion and persists for as long as the nematode feeds (17, 18). *Protocol 3* describes two simple means to provide synchronized infections for time-course studies of nematode–plant interactions. The second procedure uses plantlets and is more appropriate if the aerial parts of the plant are to be studied (19). The method based on growth-pouches offers a simple and precise procedure for the study of early events within roots (13).

Protocol 3. Infection of experimental plants

A. *Growth pouches*

1. Surface sterilize seeds of a host plant or potato tuber pieces (for PCN) in 2% sodium hypochlorite and wash extensively in sterile distilled water.

2. Germinate 1 seed/growth pouch (Northrop-King) which may be presterilized if required.

3. Germinate at 15–20°C for PCN or 25–30°C for RKN and SCN.

4. When lateral roots extend across the middle region of the pouch, cut away the plastic cover, raise a root carefully from the paper, place a 3×2 cm Whatman GF/A filter under the root (matt surface uppermost). Add 100 J2 in 5 μl of water at 0–1 cm from the root tip. Lay a second filter (matt side down) over the root and reclose the pouch with tape.

5. Remove filters and wash root after 24 h. Grow for up to 14 days.

6. Stain root tips at random to check infection. Bleach in 0.5–2% sodium hypochlorite for 1–5 min, rinse three times for 5 min in water, boil in 0.2% acid fuschin for 2 min, clear in acidified glycerol for 24 h at 60°C (29).

B. *Plantlets*

1. Establish plants in sterile white sand (BDH) in 50 ml pots with a central glassfibre wick at the optimal temperature for nematode growth.

2. 18 h before infection, place the pots on sand connected by a water column to a reservoir at 40 cm below the pot.

3. Inject about 170 J2 in 20 μl water at each of three sites around the root system. Leave for about 24 h before percolating water through the sand for several hours to wash out excess nematodes.

4. Grow plants for the infection period before excising regions of infection showing localized necrosis. Check infection of some roots as in step 6 above.

5.2.2 Root-knot nematodes

Root-knot nematodes often enter root tips and initially feed from undifferentiated cells close to the meristem. The parasite uncouples mitosis from cytokinesis in a manner that is unique to this relationship. Several multinucleate giant cells are formed with mitosis occurring asynchronously within them. Secondary vascular tissue is differentiated around the central giant cells and head of the nematode (16). The plant response involves galling in the region of the nematode with the swollen root of sufficient diameter to enclose the swollen female. The time course of events can be studied by staining nematodes within the roots (*Protocol 3*).

5.3 Plant responses

We do not know all the plant responses induced by nematode invasion but some defences recognized by the study of other host–pathogen interactions may also have a role against these animals. Non-chemical defences against nematodes are uncommon but maize prevents fertile egg formation by enclosing females of cereal cyst nematodes in relatively broad roots where males are unable to fertilize them (20). Some plants release toxins that influence nematodes in the soil environment; mustard plants suppress the hatching of cyst nematodes and at least one cultivar of asparagus releases toxins against ectoparasitic nematodes. More common defence mechanisms arise after nematode invasion and may involve hypersensitivity. Phytoalexins in some plants may have anti-nematode activity and hypersensitivity involving necrosis, lignification, and raised peroxidase activity occurs, particularly in response to migrating endoparasites (21, 22). Adverse plant responses often result in the re-emergence of the nematode from the incompatible plant. Pathogenesis-related proteins are produced both locally and systemically in response to infection of potato by potato cyst nematode (19). In addition, protease inhibitor proteins also occur in leaves and stems of the potato at this time (23). Some of these responses may be of indirect benefit in augmenting defences against other pathogens or pests during the stress induced by nematodes.

Resistance may be directed against the nematode or the cells from which the animal feeds. Resistance of the potato cultivar, Maris Piper, against *Globodera rostochiensis* pathotype Ro1 and Ro4 is expressed by increased necrosis and lignification around the syncytium which fails to contact the vascular elements and does not become a transfer cell (18, 24). This prevents the development of females but not males.

5.4 Disease associations

A few nematodes are vectors of a narrow range of plant viruses (e.g. nepoviruses by *Xiphinema* spp. and *Longidorus* spp.; and certain tobraviruses by *Trichodorus* spp.) (25). In addition, nematodes in association with fungi are transmitted by specific insect vectors and cause a few important conditions such as pine wilt disease and red ring disease of coconut (2). In restricted areas of Australia a nematode introduces a *Corynebacterium* to the seed head of rye grass which then becomes highly toxic to grazing sheep. Disease associations with both bacteria and fungi, particularly *Fusarium* spp., contribute considerably to the economic status of *Meloidogyne* spp. Beneficial nodulation of legumes by *Rhizobium* spp. can also be suppressed by soybean cyst and pea cyst nematodes (26).

6. Resistance

Plant resistance is defined by failure of nematodes to reproduce on a genotype of a host plant species. Dominant, partially dominant, and recessive modes of inheritance occur based on one to three plant genes. A gene-for-gene hypothesis has been proposed in some cases with typically a dominant plant resistance gene being countered by a recessive virulence gene in the nematode (27). The success of plant breeders and the need for continued effort can be summarized by reference to two examples described in Sections 6.1 and 6.2.

6.1 *Globodera* spp.

Different sources of resistance occur and allow subdivision of populations of potato cyst nematode in Europe into five forms of *G. rostochiensis* (Ro1–5) and three of *G. pallida* (Pa1–3). These pathotypes are defined as forms of one species that differ in reproductive success on defined host plants known to express genes for resistance (20, 28). The H1 gene conferring resistance to *G. rostochiensis* Ro1 and Ro4 is virtually qualitative and widely used commercially. Within the UK, cv Maris Piper expresses H1 and is a highly successful resistant cultivar. Unfortunately, its widespread use in Britain is correlated with an increased prevalence of *G. pallida* to which this cultivar is fully susceptible.

6.2 *Meloidogyne* spp.

Morphologically similar forms or races, occur with differential abilities to reproduce on host species (20, 27). The standard test plants are tobacco (cv NC95) and cotton (cv Deltapine) for the four races of *M. incognita* whereas the two races of *M. arenaria* are differentiated by peanut (cv Florrunner). The single, dominant gene in tobacco cv NC95 confers resistance to *M. incognita* races 1 and 3 but its cropping in USA has increased the prevalence of other root-knot nematodes particularly *M. arenaria*. Most sources of resistance are not effective against more than one species of root-knot nematode with the notable exception of the LMi gene from *Lycopersicum peruvanium* which confers resistance to many species except *M. hapla*. Another limitation of resistance genes identified in tomato, bean, and sweet potato is a temperature dependence which renders them ineffective where soil temperature exceeds 28°C (20).

7. Conclusions

Nematodes represent a major plant pathology problem. They cause fundamental changes in plant cell development that are poorly understood and host responses have yet to be adequately described. Furthermore a strong case can be presented for the potential value of novel resistant cultivars for nematode control.

References

1. Atkinson, H. J. (1986). *Nematologica.*, **31**, 632.
2. Agrios, G. N. (1988). *Plant Pathology* (3rd edn), p. 803. Academic Press, San Diego.
3. Jones, F. G. W. and Jones, M. (1974). *Pests of field crops* (3rd edn). Arnold, London.
4. Dropkin, V. (1980). *Introduction to plant nematology.* Wiley, New York.
5. Southey, J. F. (1986). *Laboratory methods for work with plant and soil nematodes.* HMSO, London.
6. Sasser, J. N. (1979). In *Root-knot nematodes* (ed. F. Lamberti and C. E. Taylor), p. 359. Academic Press, London.
7. Southey, J. F. (ed.) (1978). *Plant nematology* (3rd edn). HMSO, London.
8. Trudgill, D. L., Phillips, M. S., and Alphey, T. J. W. (1987). *Outlook on Agriculture*, **16**, 167.
9. Masamune, T., Anetai, M., Takasugi, M., and Katsui, N. (1982). *Nature*, **297**, 495.
10. Atkinson, H. J., Fowler, M., and Isaac, R. E. (1987). *Annals of Applied Biology*, **110**, 115.
11. Sasser, J. N. and Carter, C. C. (ed.) (1985). *An advanced treatise on* Meloidogyne *spp.* 2 volumes. North Carolina State University Graphics, USA.
12. Mai, W. F. and Abawi, G. S. (1987). *Annu. Rev. Phytopathol.*, **25**, 317.
13. Atkinson, H. J. and Harris, P. D. (1989). *Parasitology*, **98**, 479.
14. Wyss, U. and Zunke, U. (1986). *Revue de Nematologie*, **9**, 153.
15. Atkinson, H. J., Harris, P. D., Halk, E. J., Novtiski, C., Leighton-Sands, J., Nolan, P., *et al.* (1988). *Annals of Applied Biology*, **112**, 459.
16. Jones, M. J. K. (1981). In *Plant Parasitic Nematodes*, Vol. III (ed. B. M. Zuckerman and R. A. Rohde), p. 238. Academic Press, New York.
17. Endo, B. Y. (1986). In *Cyst nematodes* (ed. F. Lamberti and C. E. Taylor), p. 133. NATO ASI Series A: Life Sciences 121, Plenum, New York.
18. Rice, S. L., Leadbetter, B. S. C., and Stone, A. R. (1985). *Physiol. Plant Pathol.*, **27**, 219.
19. Hammond-Kosack, K. E., Atkinson, H. J., and Bowles, D. J. (1989). *Physiol. Molec. Plant Pathol.*, **35**, 495.
20. Cooke, R. C. and Evans, K. (1987). In *Principles and practise of Nematode control* (ed. R. H. Brown and B. Kerry), p. 179. Academic Press, Sydney, Australia.
21. Zacheo, G. (1986). In *Cyst nematodes* (ed. F. Lamberti and C. E. Taylor), p. 163. NATO ASI: Series A: Life Sciences 121, Plenum, New York.
22. Bowles, D. J., Hammond-Kosack, K. E., Gurr, S. J., and Atkinson, H. J. (1991). In *Biochemistry and molecular biology of plant–pathogen interactions* (ed. C. J. Smith), p. 284. Clarendon Press, Oxford.
23. Rice, S. L., Stone, A. R., and Leadbetter, B. S. C. (1987). *Physiol. Plant Pathol.*, **31**, 1.
24. Sidhu, G. S. and Webster, J. M. (1981). *Botan. Rev.*, **47**, 387.
25. Jones, F. G. W., Parrott, D. M., and Perry, J. N. (1981). In *Plant parasitic nematodes*, Vol. III (ed. B. M. Zuckerman and R. A. Rohde), p. 23. Academic Press, New York.
26. Taylor, C. E. and Robertson, W. M. (1975). In *Nematode vectors of plant viruses* (ed. F.

Lamberti, C. E. Taylor, and J. W. Seinhorst), p. 253. NATO ASI Series A: Life Sciences 2, Plenum, New York.

27. Ko, M. P., Barker, K. R., and Huang, J. S. (1984). *J. Nematol.*, **16**, 97.
28. Huang, J. S. (1985). In *An advanced treatise on* Meloidogyne *spp.*, Vol. 1, Chapter 14 (ed. J. N. Sasser and C. C. Carter), p. 165. N. Carolina State University Graphics, USA.
29. Hussey, R. S. (1985). In *An advanced treatise on* Meloidogyne *spp.*, Vol. 1, Chapter 12 (ed. J. N. Sasser and C. C. Carter), p. 143. N. Carolina State University Graphics, USA.

10

Nucleic acid isolation and hybridization techniques

SARAH JANE GURR and MICHAEL J. McPHERSON

1. Introduction

Recent advances in recombinant DNA technology provide a means to analyse host–pathogen interactions at the molecular level. In this chapter we focus on various techniques currently used in our laboratory pertinent to molecular plant pathology studies. Several of our protocols have been optimized for plant and fungal systems by adapting the methods of other workers. All procedures which are well documented or commonplace elsewhere are referenced in the text.

This chapter will cover the following:

(a) The isolation and purification of high molecular weight plant and fungal DNA for Southern analyses and genomic cloning.

(b) The isolation of intact plant and fungal RNA for Northern analyses and cDNA synthesis.

(c) Northern and Southern blot analyses.

(d) The labelling, handling, and storage of nucleic acid probes.

2. Preparation of high molecular weight DNA

2.1 Introduction

Good quality plant and fungal DNA can be prepared by a variety of methods. If the DNA is to be used for the construction of genomic libraries or in Southern blot analysis it is essential that the DNA is intact and of a high molecular weight (between 100–200 kb). Several methods published in the early 1980s focused on the isolation of nuclei prior to the extraction of DNA. This is important in the preparation of plant nuclear DNA free of chloroplast contamination. In this chapter we discuss the preparation of plant DNA following the isolation of nuclei and the preparation of fungal DNA from frozen mycelium.

The size and integrity of the plant and fungal DNA can be assessed by slow, low-voltage agarose gel electrophoresis in TBE or TAE buffer (10) or by pulsed-field gradient gel electrophoresis (1).

2.2 Harvesting plant and fungal tissue

We have found that young expanding leaves and exponentially growing fungal liquid cultures yield the greatest amounts of good quality DNA. Leaves should be harvested directly into foil packets and immediately frozen in liquid nitrogen. Mycelium is collected by filtration through several layers of muslin lining a Buchner funnel and then extensively water-washed. The mycelium is blotted dry between layers of tissue and immediately frozen in liquid nitrogen in foil packets. The tissue is gently broken into fine pieces by crushing the foil envelopes with a pestle. Care must be taken not to allow the tissue to thaw as this will cause lysis and the release of endogenous nucleases. Frozen tissue can be stored at $-70°C$ for long periods of time (≥ 1 year).

2.3 DNA extraction

2.3.1 Plant tissue

Jofuku and Goldberg (2) recently described a procedure for the preparation of high molecular weight DNA from isolated plant nuclei. This method involves the extraction and lysis of nuclei in a high pH buffer (pH 9.5) which minimizes nuclease activity in both monocotyledonous and dictoyledonous plants. We have used this method to prepare intact and high molecular weight barley, soybean, potato, tomato, and *Arabidopsis* DNA. This DNA has been variously used for Southern blot analysis, library construction following size fractionation of digested DNA fragments, and PCR (see Chapter 11).

2.3.2 Fungal mycelium

Two methods are routinely used for the preparation of fungal DNA. Both methods have been used to prepare high molecular weight DNA from a range of basidiomycete, ascomycete, deuteromycete, and oomycete fungi. The first method (*Protocol 1*) is based on an optimization of a previously described procedure (3). A second method is adequately described elsewhere (4).

Protocol 1. Preparation of fungal DNA

1. Pre-cool a pestle and mortar at 4°C. Grind 5 g mycelium to a fine powder in liquid nitrogen and transfer to a plastic Sterilin tube, ensuring that the tissue does not thaw.

2. Add 2–3 ml of ice-cold extraction buffer (0.5 M sucrose, 25 mM Tris–HCl pH 7.5, 20 mM Na_2 EDTA) per g of tissue.

3. Add lysis buffer (Sarkosyl to a final concentration of 4% and SDS to 0.5% v/v) and incubate at 60°C for 60–90 min after gently inverting the tube to form a more homogeneous solution.

Protocol 1. *Continued*

4. Transfer to centrifuge tubes (Sorvall, Oak Ridge) and spin in SS34 rotor (Sorvall), 12 000 g for 10 min at room temperature.

5. Retain the supernatant, add 10 mg/ml proteinase K to a final concentration of 200 μg/ml and incubate at 35°C for a minimum of 5 h (overnight incubation is convenient and yields good DNA).

6. Add an equal volume of precipitation buffer (30% PEG 6000, 1.5 M NaCl), gently invert and leave on ice for 60 min.

7. Transfer the DNA solution to a Corex tube and spin in SS34 rotor, 12 000 g for 20 min at room temperature (to avoid precipitation of the SDS). Discard the supernatant and resuspend the pellet in 0.5–1 ml of resuspension buffer (50 mM Tris–HCl, pH 8, 10 mM Na_2 EDTA) per g of tissue.[a]

8. Add an equal volume of SS phenol (8) and invert gently several times. Separate the phases in a bench-top centrifuge, 1000 g for 10 min.

9. Add DNAase-free RNAase A (10 mg/ml solution) to a final concentration of 50 μg/ml to the aqueous phase in a fresh tube and digest for 1 h at 37°C. Small amounts of RNA may remain; this does not interfere with restriction enzyme digestion.

10. Repeat the phenol extraction until the interface between the aqueous and phenol phases is clear. This may take 5 or more extractions.

11. Extract the aqueous phase with an equal volume of chloroform (CH_3Cl). Invert gently and spin in a bench-top centrifuge at 1000 g for 10 min. Repeat this step when preparing DNA from oomycete fungi and separate the phases by centrifuging 12 000 g for 20 min at room temperature.

12. Precipitate the DNA by adding 3 M sodium acetate (pH 6.5) to a final concentration of 0.3 M and 2 vol. of cold ethanol. Genomic DNA will precipitate immediately after the solution is mixed by gentle inversion.

13. Pellet the DNA by centrifugation at 12 000 g for 10 min at 4°C.

14. Wash the pellet in 1 ml cold 70% ethanol and repeat step 13.

15. Decant ethanol, allow pellet to air-dry, and resuspend in $1 \times$ TE buffer (10 mM Tris–HCl, pH 7.5, 1 mM Na_2 EDTA).

---

[a] Certain fungal DNA (e.g. oomycetes) forms a viscous ring around the side of the centrifuge tube rather than a pellet. The supernatant should therefore be removed with extreme care.

Genomic DNA dissolves slowly in TE following ethanol precipitation, particularly if the concentration exceeds 1 μg/μl. It may be necessary to resuspend the DNA on a rotary table at 20 r.p.m. overnight at room temperature.

3. RNA extraction

3.1 Introduction

Many methods describe the isolation of RNA from intact cells, nuclei, and polysomes. The isolation of undegraded RNA requires the inhibition of high levels of endogenous ribonucleases and the efficient deproteinization of the RNA. High levels of endogenous RNAases pose a threat to the isolation of intact RNA from plants and fungi. We therefore favour the use of guanidinium HCl (GuHCl), coupled with 2-mercaptoethanol as powerful protein denaturing and reducing agents respectively.

Recently a method was described for the fast and efficient isolation of RNA from various plant tissues based on extraction with GuHCl (5). We have optimized and abridged this protocol in order to reproducibly prepare intact RNA from up to 60 different root, stem, cotyledon, or leaf samples per day. Intact fungal RNA has also been prepared using this method. Plant and fungal RNA prepared as described has been used in Northern blot analysis, cDNA synthesis, and PCR.

3.2 Harvesting plant and fungal tissue

Plant tissue and fungal mycelium is harvested and immediately frozen in liquid nitrogen. The tissue in the foil packets is gently crushed to a fine powder with a pestle and stored at $-70°C$ as described in Section 2.2.

3.3 Isolation of total RNA

All glassware and disposable plastics are pre-treated in 0.02% diethylpyrocarbonate (DEPC) and baked at 180°C for >2 h or autoclaved respectively. Gloves must be worn at all times.

Protocol 2. Preparation of total RNA

1. Pre-cool disposable plastic 20 ml tubes (e.g.Sarstedt) in liquid nitrogen.

2. Transfer 200–400 mg pre-crushed tissue to the tubes and add 0.5 ml GuHCl buffer*a* and 0.5 ml phenol/CH_3Cl/isoamylalcohol (8).

3. Allow the tubes to thaw on ice and homogenize the samples by 15 strokes of Polytron blender (Kinematica) returning the samples to ice after grinding.

4. Separate aqueous and phenol phases by centrifugation, 1000 g at room temperature for 10 min in bench-top centrifuge.

5. Remove upper aqueous phase to microcentrifuge tubes avoiding any interface material. Re-extract with phenol, vortex briefly, and spin at 13 000 g at room temperature for 10 min in a microcentrifuge.

6. Repeat step 5 until the interface between the aqueous and the phenol phase is

112

Protocol 2. *Continued*

clean (usually 3 extractions with plant tissue and >5 times with fungal mycelium).

7. Add 0.2 vol. 1 M acetic acid and 0.7 vol. cold 96% ethanol to the aqueous phase and precipitate at $-20°C$ overnight. The precipitate may not become visible immediately.

8. Pellet the RNA by spinning at 13 000 g in a microcentrifuge for 10 min at 4°C. Wash the pellet to remove low molecular weight RNA species and polysaccharides by vortexing the pellet in 400 μl 3 M sodium acetate at pH 5.5 at 4°C.

9. Pellet the RNA and repeat step 8. Remove the salt with a final 70% cold ethanol wash.

10. Dissolve the pellet in 30–50 μl DEPC-treated water. If the pellet is slow to dissolve, heat sample to 95°C for 2 min, vortex, and ice quench.

11. Assess the quality and quantity of the prepared RNA by comparing $OD_{260}:OD_{280}$ ratios.

12. Store the RNA at $-70°C$.

^{a} GuHCl buffer: 8 M guanidinium–HCl, 20 mM MES (4-morpholino-ethanol-sulphonic acid), 20 mM Na$_2$ EDTA, adjust to pH 7 with NaOH, add 50 mM 2-mercaptoethanol prior to use.

3.4 Northern blot analysis

Several comprehensive articles review the merits of different denaturing systems for the electrophoresis of RNA. These include glyoxal, methylmercuric hydroxide, and formaldehyde gels (6).

Formaldehyde agarose gels allow good size separation and high resolution of RNA species and this is the electrophoresis system routinely used in our laboratory. Most methods include formaldehyde at a final concentration of 2.2 M. At this concentration formaldehyde-denatured nucleic acids are poorly visible following ethidium bromide staining. Furthermore, many authors recommend that if the RNA is to be blotted the gel should not be ethidium bromide stained as this interferes with transfer to the filter (7). This is not our experience. Gels with formaldehyde at a final concentration of 0.66 M and ethidium bromide 0.5 μg/ml show clear staining of the RNA species and good transfer to nylon membranes or to nitrocellulose. Furthermore, the integrity and equivalence of loading can be readily assessed using this method.

Protocol 3. Northern blot analysis

1. Electrophorese RNA samples through 0.66 M formaldehyde gels containing 0.5 μg/ml ethidium bromide at 4°C as described in ref. 8.

Protocol 3. *Continued*

2. Photograph gel and rinse twice in $10 \times$ SSC (1.5 M NaCl, 0.15 M Na$_3$ citrate) at 42°C for 10 min (to aid removal of formaldehyde from the gel).

3. Blot on to nylon membrane (e.g. Hybond N$^+$, Amersham), or when using the ECL system on to nitrocellulose (see Section 5.1.6), for at least 12 h.

4. Bake filter for 2 h at 80°C.

5. Prehybridize filters for a minimum of 6 h in 50% deionized formamide, $5 \times$ SSC, $1 \times$ Denhardts (0.02% Ficoll 400, 0.02% BSA fraction V, 0.02% PVP (polyvinyl-pyrrolidone)), 100 μg/ml sonicated herring sperm DNA (boiled for 10 min prior to use), 12.5 mM tetrasodium pyrophosphate (pH 6.5).

6. Boil ^{32}P-labelled probe for 10 min, snap cool, and add to fresh prehybridization buffer including 6% PEG 6000. Homologous hybridizations are carried out at 42°C for > 16 h; heterologous hybridizations are at 37°C for > 20 h.

7. Wash filters briefly (2 min) in $6 \times$ SSC, 0.1% SDS at hybridization temperature. Progressively wash (in 3 to 4 steps of 20 min each) at decreasing salt concentrations to $0.1 \times$ SSC, 0.1% SDS at 60°C for homologous probes and to $1 \times$ SSC, 0.1% SDS at 42°C for heterologous probes.

8. Expose washed filters to X-ray film in cassettes at -70°C. Filters can be exposed to X-ray film at any intermediate wash step, then washing can be continued as in step 7 provided the filter is kept damp.

9. Labelled probe, in hybridization fluid, can be stored at -20°C and reused by heating above its melting temperature, usually by placing at 70°C for 30 min or boiling for 5 min.

4. Southern blot analysis

Detailed procedures for the treatment of gels prior to transfer to nylon membranes or nitrocellulose are given in (9, 10). The prehybridization, hybridization, and washing of Southern blots is described in great detail in (10, 11).

5. Labelling, handling, and storage of nucleic acid probes

The detection of labelled nucleic acids forms the basis of most major, and now commonplace, molecular biological techniques including filter and *in situ* hybridization, nucleic acid sequencing, and gene and genome mapping.

The sensitivity and resolution demanded by a particular experiment determines the choice of label and labelling technique. Until recently, most nucleic acids were labelled with ^{32}P, ^{35}S, or ^{3}H. ^{32}P-labelled probes are widely used and highly sensitive (see Section 5.3). For example, a single copy gene of around 1 kb accounts for some 5 pg in a 10 μg preparation of potato DNA (genome size of approximately 2×10^9 bp). This is readily detected using ^{32}P where the lowest detection limit of a random-primer labelled probe (see Section 5.3.2) is around 0.1 pg. However, radiolabelled probes are inherently unstable and are potentially hazardous to the researcher after prolonged and regular exposure.

Several methods for the incorporation of non-radioactive reporter molecules into nucleic acids using chemical or enzymatic methods have been described (12). These include such reporter molecules as biotin (13), sulphonated cytidine (14), and acetylaminofluorenyl modified guanosine (15). Detection by antibodies or in the case of biotin by avidin or streptavidin are coupled with colorimetric, fluorometric, or chemiluminescent signal generating systems. Until recently biotin was the most widely used non-radioactive labelling system. At least two novel non-radioactive nucleic acid hybridization and detection systems have now been developed including the enhanced chemiluminescent (ECL) method of Amersham International plc (see Section 5.1) and the Digoxigenin labelling (DIG) system of Boëhringer-Mannheim (see Section 5.2).

5.1 Enhanced chemiluminescence system (ECL)

The ECL system couples the nucleic acid labelling procedure developed by Renz and Kurz (16) with the detection by enhanced light emission catalysed by horse-radish peroxidase.

5.1.1 Labelling reaction

The labelling reaction is rapid and simply consists of a 5 min DNA denaturation step at 95°C, the addition of horse-radish peroxidase and glutaraldehyde, then incubation at 37°C for 10 min. The probe can be used immediately or stored at −20°C in 50% glycerol for periods of up to one year without loss of sensitivity.

5.1.2 Prehybridization

Prehybridization of the filter(s) is in prewarmed buffer (Amersham) ECL at 42°C for 10 min. This buffer contains 6 M urea, rate and volume enhancers, membrane blockers, and enzyme stabilizers. The appropriate concentration of salt is added to the buffer in the form of SSC. The association kinetics of the buffer are approximately equivalent to a buffer containing formamide.

5.1.3 Hybridization

Remove 1–2 ml prehybridization solution and mix with the labelled DNA. Add this solution to the remaining prehybridization solution around the filter. This prevents localized probe association with the filter. Hybridize overnight at 42°C with gentle shaking.

5.1.4 Washing

The filters are washed sequentially at 42°C in the recommended wash buffers I and II which contain urea and SDS and to which SSC is added before use. We have used SSC concentrations as low as × 0.01 for the final wash in homologous hybridizations.

5.1.5 Detection

The filter is placed DNA-side up on a clean tray or sheet of Saran wrap, overlaid with detection solution (containing equal volumes of the two detection agents) and incubated at room temperature for 1 min. The filter is lifted and the corner touched on tissue to remove excess solution before covering in Saran wrap and immediately exposing to X-ray film. After exposure for 1 min the film is removed and replaced by a second film. The first film is developed and fixed in the normal manner and is used as a guide to estimate the required exposure time for the second film (10–60 min). Alternatively, following the first exposure, if the background is too high, the membranes can be rapidly placed in fresh wash solution at higher stringency and the detection step repeated.

5.1.6 Advantages and disadvantages of the ECL system

Advantages

(a) Adequate amounts of labelled probes can be stored for multiple screening.

(b) The hybridization fluids can be reused in a second round of screening.

(c) Membranes can be reprobed without removal of the previous probe. Up to seven successive reprobings have been performed without loss of sensitivity (L. Proudfoot, personal communication).

Disadvantages

(a) Comparatively high probe concentrations are needed; between 10–20 ng/ml hybridization buffer. A typical labelling reaction may therefore require 200 ng purified DNA as the probe.

(b) The probe length must be greater than 300 bp. However, Amersham have recently launched an oligonucleotide labelling and detection system based on ECL.

(c) In our experience ECL is 10 times less sensitive than ^{32}P-labelled probes.

(d) The Amersham hybridization buffer dissolves extremely slowly. White crystals of blocking agent plus urea *must* be dissolved before use, otherwise the X-rays will be of poor quality with an unacceptably dark and smeared background.

(e) The probe must be adequately denatured at 95°C for 5 min and *must* be in ≤ 10 mM salt or salt-free. High salt concentrations will abolish labelling; inadequately denatured probes will radically reduce labelling efficiency.

(f) Labelling and hybridization temperatures must be rigorously controlled; hybridization temperature must not exceed 42°C, labelling temperature must be at 37°C.

(g) The blocking agent required when using nylon membranes contains RNAases. Northern analysis must therefore be carried out using nitrocellulose membranes.

The best results are obtained with Hybond ECL nitrocellulose or Hybond N^+ membranes and Hybond ECL film (Amersham) although in our experience both Kodak XAR and Fuji X-ray film have produced acceptable results. We have used the ECL system in Northern and Southern analysis (*Figure 1*) and to screen plaque lifts.

5.2 Digoxigenin labelling system (DIG)

The DIG system is based on the random-primed (see Section 5.3.2) incorporation of digoxigenin-labelled dUTP coupled with detection of the target DNA by ELISA.

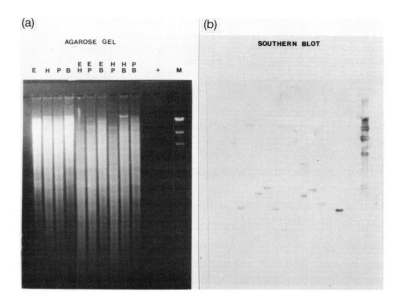

Figure 1. Southern blot analysis using ECL detection system. (a) Agarose gel of single and double restriction enzyme digests of fungal genomic DNA (E, *Eco*RI; H, *Hind*III; P, *Pst*I; B, *Bam*HI). (b) ECL film of (a). 200 ng of a 1.4 kb DNA fragment generated by PCR (Chapter 11) was labelled using the ECL system. Hybridization and washing were at 42°C. Hybridizing fragments were detected after 5 min exposure. Markers (M) were cut from gel (a) and hybridized separately to ECL labelled Lambda *Hind*III fragments.

5.2.1 Labelling

The labelling reaction is fairly fast. It consists of a 10 min DNA denaturation step at 95°C followed by the addition of hexanucleotides, dNTP label mix and Klenow polymerase followed by incubation at 37°C for more than 1 h. Greatest levels of incorporation are achieved with 5 h incubation and the probe is stable for up to 20 h. Ethanol precipitated probe can be stored at -20°C up to 1 year without loss of sensitivity.

5.2.2 Prehybridization

Filters are prehybridized, with gentle agitation, in either aqueous solution at 68°C or 50% formamide at 42°C and containing SSC, sarkosyl, SDS, and a blocking reagent (fractionated milk solids).

5.2.3 Hybridization

Hybridization is in fresh prehybridization fluid including the DIG labelled probe for a minimum of 6 h.

5.2.4 Washing

Filters are washed in SSC and SDS according to normal stringency parameters (10) and can be processed immediately or stored for later detection of hybridized DNA.

5.2.5 Detection

Immunological detection is at room temperature by incubating for 30 min in antibody-conjugate solutions followed by dark incubation in tetrazolium salts and phosphate solution. The coloured precipitate appears within minutes and is complete within one day. The membrane can be photographed directly or laser photocopied.

5.2.6 Advantages and disadvantages of the DIG system

Advantages

(a) Long term storage of labelled probes.

(b) Hybridization solutions can be reused.

(c) Membranes can be reprobed several times.

(d) Small amounts of DNA (10 ng) can be used as the labelling template; alternatively probes have also been prepared by PCR.

(e) Short probes, up to 100 bp, can be labelled.

(f) Nylon or nitrocellulose membranes can be used.

Disadvantages

(a) It is 10 times less sensitive than ^{32}P random primer-labelled probes.

(b) High backgrounds mask the results on filters where too little hybridization fluid has been used or ineffective blocking has taken place (we recommend ≥ 0.25 ml/cm^2 of fluid and the addition of 50 μg/ml ultra-sonicated and boiled herring sperm DNA).

(c) The appearance of small blue circles on the 'developed' nitrocellulose membrane results from colour precipitation where air bubbles have been trapped during colour development.

This labelling system has been used in Northern and Southern analyses, colony and plaque hybridization, RFLP analysis, and *in situ* hybridization.

The advent of the DIG system represents a marked improvement over biotin-labelled probes. It is not only more sensitive but produces significantly cleaner blots with reduced signal : noise ratio compared with biotin detection systems.

5.2.7 Other chemiluminescence detection systems

Boëhringer-Mannheim have recently introduced a chemiluminescent system for the detection of DIG-labelled nucleic acids. This system uses AMPPD (3-(2'-spiroadamantane)-4-methoxy-4 (3'-phosphoryloxy) phenyl-1,2-dioxetane) as a chemiluminescent substrate for anti-digoxigenin-AP conjugates and claims fast and sensitive detection in immunoassays, Southern, Northern, colony, and *in situ* hybridizations. Stratagene have also launched a labelling and detection system (FLASH), based on the chemiluminescent alkaline phosphatase substrate Lumigen PPD. We have no experience in the use of either system.

5.3 Radioactive labelling

Non-radioactive probing methods are desirable and have many applications. However, they are not sufficiently sensitive to allow detection of sequences at the level of a few picograms and are therefore inappropriate for the detection of single-copy plant genes or low abundance mRNA species. To achieve these levels of sensitivity, radioactive methods remain the only option. We shall concentrate on factors which affect the choice of radioactive labelling system, probe stability, and storage of labelled nucleic acids. The major disadvantages of ^{32}P-labelled probes are the associated health hazard, their inherent instability, and short half-life (14.3 days). However, an equivalent ^{35}S-labelled probe has longer storage life due to its half-life of 87.4 days and lower β-energy emission.

Popular methods for the uniform labelling of nucleic acids are nick-translation (17), random primer labelling (18, 19), and RNA polymerase-based labelling (20). There are also several methods for end-labelling nucleic acids (21–23). However, such techniques are of limited use in probing experiments and are not considered further.

5.3.1 Nick-translation

Until recently, nick-translated probes (17) were most commonly used. The amount of linear or circular double stranded template DNA required for a nick-translation reaction is 0.25 μg to 0.5 μg. The reaction achieves a maximum of around 65% incorporation yielding probes of a potential specific activity of 5×10^8 dpm/μg.

5.3.2 Random primer labelling

A significantly more efficient labelling system is the random primer approach (18, 19). A relatively high concentration of short oligonucleotides (usually hexamers) of random sequence are used as primers for template-directed DNA synthesis. The incorporation of isotope, in the form of radioactive nucleotide(s), can be highly efficient ($\geq 75\%$). Small amounts of template, for example, 25 ng of short (200 bp) DNA are sufficient to generate probes of high specific activity, 5×10^9 dpm/μg.

The highest labelling efficiencies are achieved using DNA fragments recovered from Nusieve GTG low melting temperature agarose (FMC Bioproducts) by elution on to DEAE membranes. Lower levels of incorporation can be achieved with relatively impure templates or by direct labelling of the DNA in low melting point agarose.

The importance of vector-free fragments for probe labelling in the screening of libraries constructed in the same or similar vector as the probe cannot be over-emphasized. It is essential to purify small fragments at least twice through low melting point agarose and to 'compete-out' contaminating sequences by the addition of 0.1 to 1 mg/ml of unlabelled linearized plasmid sequences.

Following random-primer labelling of fragments we routinely separate the probe from unincorporated label in order to assess the specific activity of the probe. A small amount of blue dextran and phenol red (each in SDW) is added to the labelled probe following incubation and the mixture is separated through a 1 ml G-50 Sephadex (medium grade) column in a 2 ml disposable syringe. The addition of blue dextran and phenol red acts as a visual aid to the collection of the fractions as dextran blue elutes with the 'incorporated peak' and phenol red elutes with the 'unincorporated' nucleotides. It is not strictly necessary to remove unincorporated isotope if more than 60% incorporation is achieved in the labelling reaction.

Acknowledgements

We acknowledge financial support from Enimont, AFRC, and SERC. S. J. G. holds a Royal Society University Research Fellowship.

References

1. Schwartz, D. C. and Cantor, C. R. (1984). *Cell*, **37**, 67.
2. Jofuku, K. D. and Goldberg, R. B. (1989). In *Plant molecular biology: a practical approach* (ed. C. S. Shaw), p. 37. IRL, Oxford.
3. Tilburn, J., Scazzochio, C., Taylor, G. G., Zabicky-Zissman, J. H., Lockington, R. A., and Davies, R. W. (1983). *Gene*, **26**, 205.

4. Borges, M. I., Azevedo, M. O., Bonatelli Jnr., R., Felipe, M. S., and Astolfi Fho., S. (1991). *Fungal Genetics Newsletter*. In press.
5. Logemann, J., Schell, J., and Willmitzer, L. (1987). *Anal. Biochem.*, **163**, 16.
6. Ogden, R. C. and Adams, D. A. (1987). In *Methods in enzymology*, Vol. 152 (ed. S. L. Berger and A. R. Kemnel), p. 61. Academic Press, New York.
7. Thomas, P. S. (1980). *Proc. Natl. Acad. Sci. USA*, **77**, 5201.
8. Davis, L. G., Dibner, M. D., and Battey, J. F. (1986). *Basic methods in molecular biology*. Elsevier, New York.
9. Wahl, G. M., Meinkoth, J. L., and Kimmel, A. R. (1987). In *Methods in Enzymology*, Vol. 152 (ed. S. L. Berger and A. R. Kimmel), p. 572. Academic Press, New York.
10. Maniatis, T., Fritsch, E. F., and Sambrook, J. (1982). *Molecular cloning: a laboratory manual*. Cold Spring Harbour Laboratory Press, New York.
11. Wahl, G. M. and Berger, S. L. (1987). In *Methods in enzymology*, Vol. 152 (ed. S. L. Berger and A. R. Kimmel), p. 415. Academic Press, New York.
12. Forster, A. C., McInnes, J. L., Skingle, D. C., and Symons, R. H. (1985). *Nucleic Acids Res.*, **13**, 745.
13. McInnes, J. L., Dalton, S., Vizi, P. D., and Robins, A. J. (1987). *Bio/Technology*, **15**, 269.
14. Poverenny, A. M. (1979). *Mol. Imm.*, **16**, 313.
15. Tchen, P., Fuchs, R. P. P., Sage, E., and Leng, M. (1984). *Proc. Natl. Acad. Sci. USA*, **81**, 3466.
16. Renz, M. and Kurz, C. (1984). *Nucleic Acids Res.*, **12**, 3435.
17. Rigby, P. W. J., Dieckmann, M., Rhodes, C., and Berg, P. (1977). *J. Mol. Biol.*, **113**, 237.
18. Feinberg, A. P. and Vogelstein, B. (1983). *Anal. Biochem.*, **132**, 6.
19. Feinberg, A. P. and Vogelstein, B. (1984). *Anal. Biochem.*, **137**, 266.
20. Melton, D. A., Kreig, P. A., Rebagliati, M., Maniatis, T., Zirin, K., and Green, M. (1984). *Nucleic Acids Res.*, **12**, 7035.
21. Richardson, C. C. (1981). In *The enzymes*, Vol. XIV (ed. P. D. Boyer), p. 299. Academic Press, New York.
22. Bolluna, F. J. (1974). In *The enzymes*, Vol. X (ed. P. D. Boyer), p. 2. Academic Press, New York.
23. Thein, S. L. and Wallace, R. B. (1983). In *Human genetic diseases: a practical approach* (ed. K. E. Davies), p. 33. IRL, Oxford.

11

The polymerase chain reaction

MICHAEL J. McPHERSON, RICHARD J. OLIVER, and
SARAH JANE GURR

1. Introduction

The polymerase chain reaction (PCR), now a key tool in molecular biology, provides a rapid and powerful technique for the *in vitro* amplification of DNA sequences (see refs 1–3 for reviews). The PCR is now routinely used for gene cloning, analysis, and manipulation in many molecular biology laboratories. It has also found wide application in medicine; including diagnosis of genetic disease, mutational analysis, forensic pathology, and is proving important in such programmes as the mapping and sequencing of the human genome. The PCR has great potential for the study of plant–pathogen interactions in facilitating both gene cloning and genome analysis. Indeed, in biotrophic interactions where the host–pathogen interface is restricted to a small number of specialized cells, the PCR may provide the only mechanism for the investigation of changes in gene expression.

This chapter outlines the methodology, basic requirements, and problems of contamination of the PCR and then discusses applications of the PCR in gene cloning, for the construction of PCR-directed cDNA libraries, and for genome analysis.

2. Principles of the polymerase chain reaction (PCR)

PCR uses a thermostable *Taq* DNA polymerase to synthesize DNA from oligonucleotide primers and template DNA. The template DNA may be genomic or first-strand cDNA, or cloned sequences. Primers are designed to anneal to complementary strands of the template such that DNA synthesis initiated at each primer results in replication of the region of template between the primers.

The PCR reaction proceeds in three distinct steps governed by temperature (*Figure 1*a). First the template DNA is denatured to separate the complementary strands. Second, the mix is held at an annealing temperature to allow the primers to hybridize to their complementary sequences. Presumably at this stage the *Taq* polymerase stabilizes these base-paired structures and initiates DNA synthesis. Finally the reaction is heated to the optimum temperature (*c*. 72°C) for *Taq*

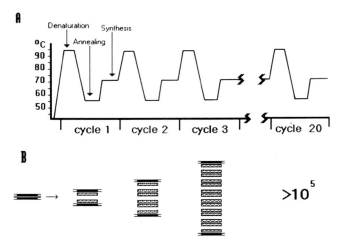

Figure 1. Schematic representation of the PCR. (A) temperature profile and (B) amplification of a target DNA.

polymerase-directed DNA synthesis. In the first cycle each template strand gives rise to a newly synthesized complement; thus the number of copies of the target region is doubled (*Figure 1*b). Similarly in each subsequent cycle there is a theoretical doubling of the DNA concentration corresponding to the target region. If a reaction occurred at 100% efficiency then some 20 cycles of PCR would produce a 10^6-fold amplification of the target DNA.

2.1 Fidelity of *Taq* polymerase

Most available *Taq* polymerases do not possess a DNA 'proof-reading' activity and are thus unable to remove bases wrongly incorporated into the nascent strand (see Section 3.2). The frequency of base misincorporation has been extensively investigated and is actually quite low if optimal conditions for the polymerization reaction are correctly established. It is suggested that error rates for base-pair changes and frameshifts are approximately 1 in 9000 and 1 in 41 000 bases synthesized, respectively (4). This mutational capacity of *Taq* polymerase should be considered during experimental design to ensure that, should they occur, mutations can be identified; especially if the prime reason for PCR amplification is to sequence the DNA. If a mutation occurs during an early cycle of PCR then it will be amplified throughout the population of molecules. It is therefore wise to sequence DNA derived from independent primary amplification reactions whether the amplified DNA is sequenced directly or is first cloned into a suitable vector. Since the mutation frequency is related to the number of bases synthesized the risk of mutation increases with the number of amplification cycles. For this reason the number of cycles of amplification should be limited to that which yields an acceptable level of product.

The importance of low level base misincorporation depends on the use to which the PCR amplified fragment is to be put. For example, there will be no effect if the PCR product is used as a probe in hybridization experiments. However, if the PCR is used to construct cDNA libraries it is important to realise that a cDNA sequence may not be accurate. The recently introduced Vent polymerase (New England Biolabs) is reported to exhibit proof reading activity and may therefore prove to be of use for high fidelity cloning applications. A comprehensive discussion of the fidelity of DNA polymerases in polymerase chain reactions can be found in ref. 5.

2.2 Contamination problems

The ability of the PCR to amplify minute amounts of template has the disadvantage that small quantities of contaminating DNA may prove to be a problem for some applications. A useful short review by Kwok and Higuchi (6) outlines contamination problems and their avoidance. To highlight the problem these authors suggest that diluting the 0.1 ml contents of a completed PCR reaction into the water of an olympic-sized swimming pool would produce a solution of which a 0.1 ml aliquot would contain some 400 molecules of DNA.

The source of contaminating material is not always obvious and one of the most likely is aerosol from pipettors. Wiping the tip of the pipettor barrels regularly with ethanol soaked tissue appears to help prevent this problem. It is important to have a designated 'clean' area for setting up PCR reactions from which other DNA samples, especially PCR products, are excluded. Solutions should be autoclaved where possible and aliquoted both for storage and to allow discard of partially used aliquots. Reactants other than DNA can be pre-mixed to reduce the number of individual pipetting steps and thus the risk of contamination.Care should be taken when removing aliquots from stock solutions of oligonucleotides to prevent contamination. Fresh autoclaved tips should be used for each dispensing step. It is equally important that non-reactants such as mineral oil do not become contaminated by careless use of a pipettor. Always pour mineral oil from the stock bottle into a microcentrifuge tube for dispensing and then discard this aliquot.

It is also important to perform control reactions in parallel with the test samples to indicate whether any contamination problems exist. At least two controls are required, a reaction containing no DNA and one containing no primers.

A recent report (7) describes the use of UV treatment of reaction components to inactivate contaminating DNA before the addition of the true template and the *Taq* polymerase. Specially designed UV irradiation apparatus, which can also be used to UV fix DNA to nylon membranes, are now available (for example the Amplirad from Genetic Research Instruments).

3. Practical requirements for the PCR

In principle the PCR is simple; DNA, primers, deoxynucleotides, buffer and *Taq* polymerase are combined in a microcentrifuge tube, overlaid with mineral oil and the tube is placed in a thermal cycler programmed to repeat a set of short incubations at pre-determined temperatures. These are typically, 95°C to denature the DNA template, annealing of primer to template at 37–60°C, depending on GC composition, and finally a DNA polymerization reaction at 72°C, the optimal temperature for *Taq* polymerase catalysed DNA synthesis. In some circumstances the annealing temperature can be as high as 72°C allowing two-step PCRs.

3.1 Thermal cycle instruments

Suitable basic cyclers are now available from a range of suppliers and require only simple programming. Such instruments will cycle through a series of pre-set temperatures a pre-defined number of times and will often hold the samples at 4–6°C on completion. This latter facility is quite useful if overnight reactions are anticipated. More sophisticated instruments allow the rate of heating and cooling (i.e. ramping; secs/°C) to be regulated and may also provide additional features such as pre-programmed pauses to allow sampling or addition of further reactants. Most of the experiments discussed in this chapter were performed with PREM (LEP) or PTC (M. J. Research) instruments.

3.2 Thermostable *Taq* polymerases

Taq polymerases are now available from a range of suppliers. Generally the naturally produced enzyme has been superseded by cloned versions. The experiments described in this chapter used *Taq* polymerase from Cetus (AmpliTaq), Cambio Type III (uncloned), Promega, or Northumbrian Biological Ltd. Since there is little difference between the quality of *Taq* polymerases it is often best to try the cheapest first. One exception is the new Vent polymerase from New England Biolabs which is reported to have proof-reading activity and which may therefore prove important for certain direct cloning applications.

3.3 PCR buffers

Most suppliers of *Taq* polymerase provide reaction buffer with the enzyme. Otherwise the following general buffer produces good results with all *Taq* polymerases examined (G. R. Taylor, personal communication).

Standard PCR buffer (10 ×)
100 mM Tris–HCl (pH 8.3, 25°C), 500 mM KCl, 15 mM $MgCl_2$, 1 mg/ml gelatin, 0.1% Tween-20, 0.1% NP-40.

This solution can be autoclaved prior to addition of the non-ionic detergents then aliquoted and stored at $-20°C$.

3.4 dNTPs

Solutions can be purchased from Pharmacia or lyophilized preparations can be redissolved and neutralized. Conveniently 100 mM stock solutions are stored at $-70°C$. Prepare working solutions by diluting stocks to 2 mM in sterile double distilled water. Typically 200 μl aliquots of a working stock are prepared by adding 4 μl of each dNTP stock to 186 μl water. This solution may be stored for several weeks at $-20°C$.

3.5 Oligonucleotide primers

Oligonucleotide stocks are stored at $-20°C$ in ammonia and are not routinely purified following deprotection. Usually an aliquot of this stock containing several nmoles of oligonucleotide is dried under vacuum and redissolved at 100 pmole/μl in double distilled water. These solutions may be stored for several weeks at $-20°C$. Alternatively aliquots of the stock may be dried down directly in the PCR reaction tube immediately prior to use by placing at 95°C to drive off the ammonium. We prefer the former approach for most applications since there is less chance of contaminating the stock ammonia solution (see Section 2.2). The design of primers for the PCR is discussed in Section 5.

3.6 Template DNA

3.6.1 Genomic DNA

Various methods have been used to prepare genomic DNA from bacteria, fungi, plants, and animals. It may be necessary to purify plant DNA by CsCl gradient centrifugation (see Chapter 10); bacterial, fungal, and animal DNA do not require this step. Dissolve the DNA in double distilled water at a concentration of around 0.1 mg/ml.

Bacteria
DNA from both Gram-negative and Gram-positive strains can be prepared by the same method (8). Usually it is sufficient to prepare DNA on a small-scale according to a modification of this method (9).

Filamentous fungi
The method described in Chapter 10 of this book and the spermidine buffer method of Borges *et al.* (10) both work well.

Plant tissue
DNA from a range of plants including tomato, potato, *Arabidopsis*, soybean, and barley have been prepared according to the methods discussed in Chapter 10.

Animal tissue

Proteinase K-SDS methods such as that described in ref. 11 provide good quality DNA.

3.6.2 RNA isolation and first strand cDNA preparation

Methods for the isolation of plant total RNA are described in ref. 33. For the preparation of cDNA libraries we do not purify poly-A$^+$ RNA. First strand cDNA is prepared essentially according to ref. 12 using 1 μg total RNA. In our experience the first strand cDNA synthesis is the most critical step for generation of good libraries. It is worth assessing the integrity of this first-strand material by incorporating ^{32}P dCTP during synthesis then fractionating samples on a 1.4% agarose alkaline gel. Products up to 2.5 to 3 kb should be visible.

3.6.3 Plasmid and bacteriophage DNA

DNA prepared by almost any method from rapid minipreps to purification through CsCl gradients (11) may be used as template for PCR experiments. The concentration of template DNA added to the PCR varies according to the application and can be as high as 1 μg in certain mutagenesis experiments (13). More usually 0.01 to 1 ng of template is used in the PCR. As discussed in Section 4.10, clones can be rapidly screened by direct amplification from colonies or plaques without first purifying DNA.

4. Basic PCR procedures

Protocol 1 provides a method routinely used for amplification of DNA from a wide range of template including genomic DNAs (*Figure 2*), first-strand cDNA, (*Figure 7*) and plasmid or phage DNA (*Figure 4*).

Protocol 1. PCR amplification of genomic DNA

1. For a single 50 μl PCR reaction add the following components to a 0.5 ml sterile microcentrifuge tube:

- 5 μl 10 × PCR buffer[a]
- 5 μl 2 mM dNTP solution[a]
- 1 μl of each primer (100 pmol/μl)
- X μl of double distilled water (where $X = 50 -$ sum of all other components)
- 0.1 to 0.5 μg genomic DNA or 0.01 to 0.1 μg cDNA or 0.01 ng to 0.1 μg plasmid or phage
- 2 units *Taq* polymerase[b]

2. Spin briefly (1 sec) in a microfuge to mix. Use either a special 0.5 ml tube rotor

Protocol 1. *Continued*

for the microcentrifuge or place the 0.5 ml tubes in capless 1.5 ml microcentrifuge tubes.

3. Overlay with 50 μl mineral oil to prevent evaporation during heating cycles.
4. Place tube in thermal cycler and perform PCR cycles.
5. Typical PCR reaction conditions are:
   - 95°C, 5 min
   - 25 to 45 cycles of (95°C, 1 min; 55°C, 1 min; 72°C, 1 to 5 min)
   - 55°C, 2 min
   - refrigerate until required

[a] 10 × PCR buffer and dNTPs see Sections 3.3 and 3.4 respectively.
[b] *Taq* polymerase may be added after all the other reactants by pipetting under the mineral oil. This allows addition of the enzyme towards the end of the initial (5 min) denaturation step. In practice it seems to make little difference at which point the enzyme is added and for convenience it is therefore usually added with the other reactants.

4.1 Reaction conditions

When starting to study a new system it may be useful to investigate different reaction conditions and extension times to achieve maximal efficiency.

4.1.1 Annealing temperature

The annealing temperature must be determined empirically for a particular PCR application and can range between 35°C and 72°C. We find 55°C works well for many applications. If the annealing temperature is too low non-specific priming can occur. If the annealing temperature is too high no products are usually formed.

4.1.2 Extension time and number of cycles

The extension rate of *Taq* polymerase is > 60 nucleotides per sec at 70°C (14). A one minute extension at 72°C should therefore result in the incorporation of at least 3600 nucleotides. In fact the minimum extension time we use, even for small fragments of around 100 nucleotides, is one minute.

Template quality and quantity and the efficiency of the PCR govern the number of cycles required to generate sufficient product. For example with plasmid or phage DNA, 20 to 25 cycles is usually sufficient whereas with genomic DNA at least 30 and more usually 35 to 45 cycles may be necessary. In a preliminary experiment, aliquots of the reaction mix may be withdrawn every five cycles, starting at 20 cycles. Analysis of the samples by agarose gel electrophoresis (Section 4.4) can provide a useful indicator of the optimal number of cycles required to generate sufficient product. It is best to perform the minimum number of cycles necessary to minimize potential accumulation of *Taq* induced errors (Section 2.1).

4.1.3 Magnesium concentration

Occassionally a reaction will not proceed well under standard conditions using the standard reaction buffer (p. 126). This may reflect a different requirement for Mg^{2+} ions. A series of parallel amplifications using standard buffer containing a range of $MgCl_2$ concentrations from 1 to 5 mM will usually reveal the optimum Mg^{2+} conditions for a given template/primers combination.

4.2 Multiple PCR reactions

Where possible a pre-mix containing common reactants should be prepared. This reduces the number of repetitive pipetting steps and thus potential cross contamination (Section 2.2). Calculate the required amounts of each reaction component and multiply by the number of reactions to be set up. To allow for minor pipetting errors we add sufficient reactants for at least one additional tube (e.g. if 5 reactions are being set up then sufficient reactants for 6 reactions are pre-mixed). Aliquot this pre-mix into reaction tubes prior to the addition of other reactants.

4.3 Specificity of the PCR

In most cases it is possible to determine conditions (of annealing temperature, number of cycles, and Mg^{2+} concentration) which produce a unique PCR amplification product. However, where the GC composition of template DNA is high, non-specific annealing of the primers can occur leading to multiple banding patterns. Specificity of priming in such cases can be improved by the incorporation of formamide (15) or tetramethylammonium chloride (16).

4.4 Analysis of PCR products

Routine analysis of PCR products is by agarose gel electrophoresis through a gel of appropriate quality and concentration (see *Figure 2*). As most PCR products are relatively short, <2 kb, gels usually contain 2 to 4% standard (e.g. Sigma type II) or a high quality, low melting point (LMP) agarose (e.g. Nusieve GTG, FMC Bioproducts). The latter is used where a PCR product is to be recovered from the gel for further manipulation (Section 4.6). Agarose gel casting and handling procedures are outlined in *Protocol 2*. Preparative gels may be prepared in the same manner and the PCR reaction loaded following sample concentration by either ethanol precipitation or other appropriate method such as Geneclean (Bio 101 Inc.) or by extraction with butanol (11).

Protocol 2. Agarose gel analysis of PCR products

1. Melt 0.5 to 2 g of agarose in 50 ml TAE buffer[a] (to give a 1 to 4% gel), this may be conveniently carried out in a microwave oven. Allow the agarose solution

Protocol 2. *Continued*

to cool to around 50°C in a water bath and add 5 μl of 10 mg/ml ethidium bromide. *Caution—ethidium bromide is carcinogenic so always wear gloves when handling solutions or gels.*

2. Cast a mini-gel by pouring the agarose solution into a suitable gel mould with a suitable well-forming comb. A 5 by 7.5 cm gel will require 12 to 15 ml agarose and can be used to separate up to 15 samples.

3. Withdraw a 5 to 10 μl sample of the PCR reaction from below the mineral oil. Take care not to withdraw any oil; expel the last of the air in the pipettor only when the tip is in the aqueous solution to clear any oil from the tip.

4. Store the remaining samples at 4°C or frozen either under the mineral oil or after transferring to a fresh tube.

5. Mix the PCR sample with 2 μl loading buffer (50% glycerol, 0.03% bromophenol blue) in a fresh microcentrifuge tube.

6. Once the agarose gel has set, remove the slot former and submerge the gel in a flat bed tank containing 1 × TAE so that the gel is covered by 2 mm of buffer.

7. Load samples and suitable molecular weight markers such as a φX174 Hae III digest then electrophorese at 3 to 4 V/cm for 30 to 60 min. Often PCR fragments will run ahead of the bromophenol marker dye so check when the marker dye has migrated about one half of the gel length.

8. Visualize the bands on a UV transilluminator and photograph if required.

ᵃ 50 × TAE stock solution is 242 g Tris-base, 57.1 ml glacial acetic acid, and 37.2 g Na₂EDTA.2H₂O per litre.

---

m a b m

Figure 2. PCR amplification of genomic DNA. PCR amplifications were performed as described in *Protocol 1* for 40 cycles and 10% of the sample analysed by electrophoresis through a 0.8% agarose gel. (a) control amplification of a 750 bp region of the galactose oxidase gene of the fungus *Dactylium dendroides*. The primers were DNA sequencing primers of 16 and 20 nt. (b) amplification of a 1.25 kb fragment of the tyrosinase gene from *N. crassa*. The primers were designed from peptide sequence data (see *Figure 6*). (m) molecular size markers of lambda *Hind* III and φX174 *Hae* III digests. The arrow indicates the position of a diffuse primer band. Figure produced by Kanjula Senvratne.

Nusieve gels are quite brittle and have a tendency to tear easily during handling. To minimize such problems it can be helpful to pour the gel in a cold room and allow to chill for about 15 min once set. To prevent tearing of the wells, the comb should be set at least 1 mm above the glass plate and once set the gel should be flooded with 1 × TAE buffer before *gently* removing the comb. It is possible to improve gel strength by preparing composite gels comprising 1% standard agarose plus an appropriate concentration of Nusieve agarose. However for preparative applications we prefer gels containing only Nusieve agarose.

4.5 Primer and primer–dimer bands

Often a small band (> 100 bp) is observed in agarose gels of PCR products and usually represents the amplification of dimerized primers. Such bands occur more frequently in PCRs where no real product is formed and the primers therefore represent potential targets for amplification. A more diffuse band representing non-dimerized products can also sometimes be seen (see *Figure 2*).

4.6 Recovery of DNA from agarose

For most applications we prefer to recover a PCR product as a defined band from a gel rather than to use an unpurified PCR reaction mix for further manipulations. A Nusieve or similar quality gel should be used to separate the PCR products and the DNA containing band, visualized under UV illumination, then excised with a razor blade or scalpel. Minimize UV exposure to prevent irradiation damage to the DNA.

When multiple samples are separated on the same gel, potential cross contamination can be minimized by leaving empty wells between sample wells and by using fresh blades to excise bands. Two rapid methods for recovering DNA from gel slices are described in *Protocol 3*. A high yield of very clean DNA may also be achieved by using GeneClean or Mermaid kits (Bio101 Inc.) according to the manufacturer's instructions.

Protocol 3. Recovery of DNA from agarose gel

A. *Spin-X filter*

1. Place the agarose gel slice in the filter unit of the Spin-X filter unit (Costar). A cheaper home-made alternative filter unit is given in ref. 13.

2. Place the tube in a − 20°C freezer until the agarose freezes, then allow the gel to thaw at room temperature. This step can be omitted but does appear to improve product recovery.

3. Centrifuge the Spin-X tube in a microcentrifuge for 5 min at *c*. 13 000 *g*.

4. Discard the filter unit containing the agarose. The solution in the tube contains the DNA which can be used with no further purification.

Protocol 3. *Continued*

B. *Dilution of agarose*

1. Place the DNA band from the agarose gel in a microcentrifuge tube.
2. Melt the agarose by incubation at 68°C for 10 min, check that all the agarose has melted, and measure the volume of solution with a pipettor.
3. Add between 2 and 9 vol. of double distilled water pre-equilibrated at 37°C. This DNA solution can be used directly for most purposes.
4. *Optional step.* Extract twice with phenol, once with chloroform and ethanol precipitate.

4.7 Reamplification of PCR products

Where the amount of DNA generated in a PCR is insufficient for subsequent manipulations, a small aliquot (*c.* 1/100) of the PCR mix may be used as template in a further PCR to produce product. DNA recovered from a gel band also provides a good and well defined substrate for reamplification.

4.8 Sequencing of PCR amplified DNA

Double-stranded PCR DNA purified by Spin-X or GeneClean methods (Section 4.6) provides good substrate for direct sequencing (*Figure 3*) using appropriate primers which may either be one of the PCR primers used for amplification, or may be a nested primer complementary to some region within the amplified sequence. PCR amplified DNA suitable for sequencing may also be purified using a Centricon-100 ultrafiltration unit (Amicon).

DNA sequencing reactions are performed using commercially available kits of Sequenase (US Biochemicals) or T7 polymerase (Pharmacia). The inclusion of non-ionic detergents NP-40 and/or Tween 20 at 0.5% improves the quality of the sequencing data (17). It has also been reported that DNA may be sequenced in melted Nusieve gel (18) and that inclusion of DMSO in the sequencing reaction improve data quality (19). We routinely sequence double stranded DNA according to *Protocol 4*, with typical results shown in *Figure 3*. However, a number of methods also exist for the generation of single-stranded PCR products suitable for sequencing, including asymmetric amplification (20), exonuclease digestion of one strand (21), or specific priming on one strand by incorporation of a priming site as part of a PCR primer (22).

4.8.1 Sequencing close to the primer

It is occasionally desirable to read DNA sequence adjacent to the sequencing primer, for example to confirm the identity of a DNA fragment by comparison with peptide sequence data from which a PCR primer was designed (see Section 5). This can usually be achieved by a combination of two features:

ACGT

Figure 3. DNA sequence analysis of PCR amplified DNA. The DNA was recovered from an agarose band using a Spin-X tube according to *Protocol 2* and was sequenced according to *Protocol 3.* Figure produced by Adam Corner.

(a) using a greater than normal dilution of the labelling mix, in the case of Sequenase a 1:20 rather than a 1:5 dilution;

(b) inclusion of Mn^{2+} in the reaction buffer (Sequenase, USB).

Protocol 4. DNA sequencing of double strand PCR DNA using Sequenase

1. Mix 0.2 to 2 µg (usually one gel band) PCR DNA recovered from agarose with 2 pmol primer in 8 µl water.
2. Heat in a boiling water bath for 5 min then freeze on dry-ice. If dry-ice is not available, rapidly chill the sample on wet ice for 5 min.
3. Thaw the solution at room temperature, add 2 µl of 10 × sequencing reaction buffer and incubate at room temperature for 20 min to allow annealing.
4. Perform sequencing reactions according to the manufacturer's instructions beginning at the labelling reaction step. If required use a 1:20 dilution of labelling mix and include manganese buffer to obtain sequence data close to the primer.

4.9 Cloning of PCR products

Theoretically the cloning of PCR products should be as straightforward as cloning any other DNA molecule and is often facilitated by including restriction sites at the 5'-ends of PCR primers (see Section 5). Occasionally, however, PCR-derived fragments have proved difficult to clone and even purification through Nusieve gels or by phenol extraction does not help. To circumvent such problems either: (a) incorporate a proteinase K digestion (50 μg/ml) step prior to restriction enzyme digestion (23) and gel purification; or (b) directly ligate PCR products into a T-vector pCR™ 1000 (TA Cloning System, Invitrogen (24, 25)). These cloning systems exploit the non-template-dependent activity of *Taq* polymerases during the PCR that results in addition of single A residues to the 3'-end of nascent strands. The vectors provide complementary T-overhangs to which the PCR products can be directly ligated.

4.10 PCR screen of recombinant clones

As mentioned in Section 3.2.7, the PCR can be used to rapidly screen recombinant clones for the presence of insert DNA (*Figure 4*) and even for the relative orientation of the insert. Both bacterial colonies and bacteriophage plaques provide sufficient template DNA for such analysis and an efficient method for amplification of pUC and M13 inserts, described by Gussow and Clackson (26), is provided in *Protocol 5*. We also routinely use this method to screen for inserts in lambda vectors such as λZAP (see *Figure 4*). The following primer pairs work well for the indicated vectors:

(a) For M13, pUC, pBluescript, λZAP

   5'-GTAAAACGACGGCCAGT-3'
   5'-AAACAGCTATGACCATG-3'

(b) For pBluescript, λZAP and other vector with T7 and T3 promoter sequences

   T7 5'-AATACGACTCACTATAG-3'
   T3 5'-ATTAACCCTCACTAAAG-3'

(c) for λgt11

   5'-GGTGGCGACGACTCCTGGAGCC-3'
   5'-GACACCAGACCAACTGGTAATG-3'

Protocol 5. Rapid PCR screen of recombinant clones

1. Prepare a PCR mix of 20 μl per sample plus at least one additional aliquot to account for pipetting errors.

   - double distilled water to 20 μl
   - 10 × PCR buffer[a]   2 μl
   - 2.5 mM dNTPs   2 μl

135

Protocol 5. *Continued*

- each primer   10 pmol
- *Taq* polymerase   1 unit

2. Aliquot 20 μl of PCR mix into 0.5 ml microcentrifuge tubes.

3. Toothpick individual plaques or a very small part of a colony into the PCR mix. If required, transfer toothpick to growth media, buffer, or agar plate to recover phage or bacteria.

4. Overlay with 20 μl light mineral oil and perform 30 to 35 cycles of 94°C, 1 min; 55°C, 1 min; 72°C, 2 min then 5 min at 60°C.

5. Analyse 5 μl aliquots by agarose gel electrophoresis (*Protocol 2*).

---

[a] 10 × PCR buffer, see section 3.3.

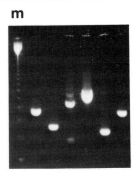

Figure 4. Amplification of DNA from plaques. Lambda ZAP plaques were picked directly into PCR reaction buffer and amplified according to *Protocol 5* using M13 forward and reverse primers. (m) 123 bp markers (Pharmacia).

4.10.1 Alternative rapid PCR screen of lambda plaques

An alternative method to that in *Protocol 5* for analysis of lambda clones involves the normal harvest of plaques into 0.5 ml SM buffer (0.1 M NaCl, 10 mM $MgCl_2$, 50 mM Tris–HCl, pH 7.5, 0.01% gelatin) containing 20 μl chloroform. Vortex and allow the phage to diffuse from the agar plug for around 30 min at room temperature, spin in a microcentrifuge for 30 sec. Use 10 μl of the SM buffer in place of the template DNA in *Protocol 1*, perform 35 cycles, and then analyse 10 μl on an agarose gel. The remaining SM buffer provides a stock of phage for further PCRs or for normal phage lambda manipulations.

5. Design of primers for the PCR

A PCR primer may be considered to comprise two regions, a 3′ (priming) region and a 5′ (variable) region. The most important in determining efficiency of annealing and subsequent DNA synthesis during the PCR is the 3' region which should be perfectly complementary to the template sequence. Where the DNA sequence is known the selection of a suitable primer sequence is straightforward. The priming region should normally be 20 to 25 bases long although for certain applications, in particular amplification from cloned DNA, shorter (16 to 17 nt) sequencing primers can be used (see Section 4.10).

Once the 3' region of the primer has been selected, the 5' region may be designed to incorporate sequence features of use for subsequent manipulations of the amplified DNA. For example restriction enzyme sites may be included to facilitate subsequent cloning, or tails representing promoter sequences (22), specific priming sites, e.g. an M13 universal priming site, or regions complementary to a second gene to allow subsequent *in vitro* recombination (13). The length of additional 5' sequence is limited by the efficiency of oligonucleotide synthesis and subsequent purification. Currently sequences greater than 100 nt are not particularly feasible.

Where a restriction site is incorporated it is important to realize that many restriction enzymes will not cut efficiently at the extreme end of a DNA molecule. It is prudent to include several (> 4) bases 5' to the restriction site. Some enzymes, for example *Xho* I require more than ten nucleotides 5' to the restriction site (13) while others, e.g. *Not* I, require none (see Section 6.1).

A restriction site can also sometimes be incorporated within a primer by selecting a sequence where one or two base changes would create a new restriction without interfering with either priming ability or some inherent functional aspect of the DNA sequence. Since the primer is incorporated into every molecule synthesized the new restriction site will also be incorporated.

It is also important to check primer sequences to ensure they do not contain complementary regions that could allow primer association thus reducing the efficiency of the sequence as a PCR primer.

5.1 PCR primers based on peptide sequence data

The PCR is of major importance in amplifying previously uncloned DNA sequences. In particular, where a protein has been isolated, even in small quantities, it is possible to determine peptide sequence information for the design of PCR primers. This strategy may prove particularly useful for the isolation of genes encoding novel plant or pathogen proteins resulting from a host-pathogen interaction. The protein is usually subjected to proteolytic cleavage either by a chemical such as cyanogen bromide or by a specific protease such as V8 (27) and the peptide fragments separated through an acrylamide gel. After brief staining, peptides are recovered by electroelution or by transfer to a nylon membrane which is then used directly for sequencing the bound peptide. It is most convenient if two regions of peptide sequence can be determined, one preferably being the N-terminus of the protein, to allow the unambiguous relative orientation of the two PCR primers to be defined (*Figure 5*). If the N-terminus is

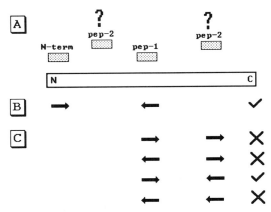

Figure 5. Illustration of the problem of primer design from peptide sequence data. (A) Regions of peptide sequence data (pep-1 and pep-2) can be positioned relative to the N-terminus allowing the unambiguous design of correctly oriented primers as shown in (B). In the absence of N-terminal data or a peptide map, the relative positions of pep-1 and pep-2 within the protein cannot be determined. As shown in (C) it may therefore be necessary to produce four primers in order to achieve DNA amplification. √ indicates successful amplification; × indicates no amplification.

blocked it is usually not possible to define the relative position of two internal regions of peptide sequence perhaps necessitating the synthesis of two primers corresponding to each region (*Figure 5c*).

5.2 Design of degenerate primers

Due to the redundancy of the genetic code a peptide sequence may be encoded by many possible nucleotide sequences depending on the amino acid composition of the peptide. For example leucine, serine, and arginine are encoded by six codons, most amino acids are encoded by two, three, or four codons with only methionione and tryptophan encoded by unique codons. For an efficient PCR it is necessary to prepare a mixture of oligonucleotides corresponding to all the possible peptide coding sequences. The synthesis of such mixtures is, however, straightforward with automated oligonucleotide synthesis instruments. *Figure 6* illustrates the design, from peptide sequence data, of redundant oligonucleotides used to amplify a fungal gene encoding the enzyme tyrosinase. At positions in the sequence where any of the four nucleotides may be present it is often more convenient to incorporate the 'universal base' inosine (28). However, it is important not to incorporate inosine as one of the three 3' nucleotides as it has been shown that these three positions must match the template sequence perfectly (29). Similar rules may be applied to the design of primers based on multiple sequence alignment data where the intention is to design "universal' primers for the amplification of gene families or the same gene from a range of organisms (30).

Michael J. McPherson, Richard J. Oliver, and Sarah Jane Gurr

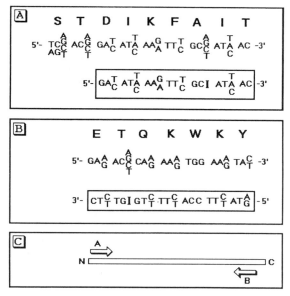

Figure 6. Design of degenerate primers from peptide sequence data. Two regions of peptide sequence data are shown for the fungal enzyme tyrosinase (34). Amino acids, shown as single letter code, have been 'back-translated' to generate the possible encoding nucleotide sequences. Inosine (I) has been incorporated at two positions of four-fold redundancy. Regions prevalent in amino acids encoded by only two codons were selected in this case, with residue such as the serine shown in (A) excluded since it may be encoded by one of six codons. Primer sequences used in the experiment illustrated in *Figure 2* are boxed. (A) N-terminal sequence back-translated for the design of a 'forward' primer (redundancy=432). (B) C-terminal sequence back-translated for the design of a 'reverse' primer (redundancy=128) which is the complement of the coding strand. (C) Relative orientations of primers, (A) and (B).

Highly degenerate PCR primers are used to amplify DNA according to *Protocol 1*. The resulting DNA is part of the target gene and can therefore be used as a homologous hybridization probe for Southern and Northern blots and for screening gene libraries under stringent conditions.

5.3 Comparison of oligonucleotide probes with PCR primers

In oligonucleotide probing experiments it is important to minimize the number of oligonucleotides in the probe mixture due to kinetics of the hybridization reaction and the relatively low level of radiolabel associated with the hybridizing proportion of the probe.

The redundancy of the probe is therefore a critical factor in determining the success of the experiment and should be kept as low as possible often restricting the regions of peptide available for probe design.

By contrast the PCR is able to cope with highly complex mixtures of primer sequences. Essentially there is no effective limit on the potential redundancy of a PCR primer and therefore no constraint on the region of peptide sequence used for the primer design. We have used primers ranging in potential redundancies from 64-fold to >73 000-fold (30). However, the more complex the primer mixture the greater the possibility of non-specific amplifications although this is easily resolved by a nested PCR strategy leading to generation of a single specific product (30).

6. cDNA cloning from small quantities of plant material

A major problem in molecular plant pathology is often the small biomass of specialized and differentiated cells at the host–pathogen interface. Analysis of gene expression within such material is now possible by using the PCR to amplify the low levels of mRNA isolated from small quantities of tissue. Methods for amplification of total mRNA populations from small numbers of mammalian cells by the use of general 3' and 5' primers have been reported (31, 32).

First strand cDNA synthesis from total (rather than purified polyA) RNA is directed by a general oligo-dT primer complementary to the poly-A tail of the mRNAs. A homopolymer tail is added to the 3' end of this first strand cDNA by incubation with terminal transferase and a single nucleotide often dGTP. These 3'-tailed cDNAs then act as substrates for PCR amplification directed by oligo-dC and oligo-dT based primers. In fact these PCR primers contain additional 5' sequences. The T primer in particular requires a relatively GC-rich tail to confer stability on the primer–template hybrid during the annealing steps that are generally performed around 50°C. Both primers contain suitable restriction sites to facilitate efficient ligation to the chosen cloning vector.

6.1 Plant PCR libraries

We have modified methods originally developed for mammalian cell cDNA library constructions for use on plant material. *Protocol 6* is intended to allow the PCR-mediated amplification of first strand cDNA products generated by any suitable method such as that outlined in Section 3.6.2. The primers used for PCR amplification are:

> 5'-GCGGCCGCTTTTTTTTTTTTTTTTTTT-3' (*Not*I)
> 5'-AAGGAATTCCCCCCCCCCCCCCCC-3' (*Eco*RI)

and to date transcripts up to 2.5 kb have been positively identified by hybridization of PCR amplified material. Libraries have been created by cloning *Eco*RI/*Not*I cut PCR products into lambda ZAP. We (S.J.G. and M.J.M.) have taken considerable care to show that such PCR-based cDNA libraries are representative. *Figure 7* shows the products of first strand cDNA synthesis and PCR amplification of RNA isolated from lettuce cotyledons infected with *Bremia lactuca*.

Michael J. McPherson, Richard J. Oliver, and Sarah Jane Gurr

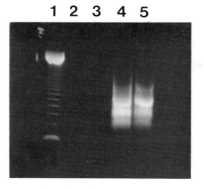

Figure 7. PCR amplification of cDNA. Lane 1, 123 bp markers (Pharmacia); lane 2, no cDNA; lane 3, no primers; lanes 4 and 5, amplification products from 1 ng aliquots of homopolymer-tails first-strand cDNA from lettuce cotyledons infected with *Bremia lectucae*.

More detailed accounts of the application of PCR for the amplification of low levels of mRNA from limited amounts of biological material is provided in ref. 33.

Protocol 6. Amplification of first-strand cDNA

1. For a single 100 μl reaction, combine the following components in a 0.5 ml sterile microcentrifuge tube:
   - 10 μl 10 × PCR buffer[a]
   - 10 μl 5 mM dNTP solution
   - 1 μl 100 pmol/μl oligo (dC) primer [*Eco*RI]
   - 1 μl 100 pmol/μl oligo (dT) primer [*Not*I]
   - X μl double distilled water (where $X = 100 -$ vol. of other components)
   - 10 pg–10 ng oligo (dG)-tailed first strand cDNA
   - 2 units *Taq* polymerase.
2. Spin briefly in a microcentrifuge and overlay with 100 μl mineral oil.
3. Place tube in thermal cycler and perform PCR reaction under the following regime:
   - 95°C, 5 min
   - cycle A (95°C, 2 min; 58°C, 2 min; 72°C, 3 min)
   - cycle B (95°C, 2 min; 40°C, 2 min; 72°C, 3 min)
   - 15–30 × cycle A
   - refrigerate
4. Split the reaction mix into four 25 μl aliquots in fresh 0.5 ml microcentrifuge

141

Protocol 4. *Continued*

    tubes. Add fresh $10 \times$ buffer, dNTPs, primers, and *Taq* polymerase to bring the volume of each aliquot to 100 μl.

5. Reamplify the aliquots for $5 \times$ cycles A (step 3).

---

a $10 \times$ PCR buffer: 100 mM Tris–HCl, pH 8.3 (25°C), 500 mM KCl, 40 mM MgCl$_2$, 0.1% gelatin.

The amplified fractions can be recovered from agarose gels (Section 4.6) following restriction enzyme digestion and can be cloned into a suitably cleaved vector.

7. Genome analysis by RAPD PCR

The PCR may also be used as a diagnostic tool and of great promise for molecular plant pathologists is the technique known as RAPD PCR. Considerable effort has been invested in the development and use of RFLP technology to map the genomes of many diverse organisms. The processes of probe isolation and labelling, DNA preparation, restriction digestion and Southern blotting, hybridization and autoradiography are laborious. Recently, however, a group at DuPont (USA) have adapted the PCR to rapidly achieve the same aims as RFLP technology. This technique, known as RAPD (Random Amplified Polymorphic DNA) PCR, has manifest applications in genetic analysis, in identification of species, and in taxonomic studies. The amplified DNA can be cloned and used as a starting point for chromosomal walks. Also, the relationship between isolates or pathovars can be assessed by comparing shared and unique bands produced by a range of primers. The technique still remains to be widely tested under a variety of conditions in order to assess its robustness and general usefulness but it is already proving to be an exciting technology with significant promise and is discussed further in Volume II, Chapter 13.

7.1 Principles of RAPD PCR

The key to RAPD PCR is that oligonucleotides of arbitrary sequence are used as PCR primers and if these primers are short, then complementary sequences will occur frequently in the target genome. There is therefore a finite chance that pairs of sequences complementary to the primer will not only be close to one another but will also be arranged with 3' ends pointing towards each other. Under these conditions, annealing of the primer to the target genome will result in the production of an amplified fragment after appropriate thermal cycling.

7.2 RAPD amplification of genomic DNA

In practice, 9 and 10-mer primers have been successfully used at an annealing temperature (*Protocol 7*, step 1) of 35°C. In principle shorter primers could be used although this would require even lower annealing temperatures. Primers used according to *Protocol 7* have all been 10-mers of which, five to eight bases were G or C and primers were used without further purification (Section 3.5).

As in RFLP analysis, the arbitrary PCR primers may reveal polymorphisms between isolates of a species, and between species. These polymorphisms can be followed in crosses to establish genetic maps. *Protocol 7* describes the technique used for a preliminary evaluation of RAPD PCR for studying polymorphisms in *Betula* spp. and *Cladosporium fulvum*.

Protocol 7. RAPD PCR with plant and fungal DNA

1. Combine in a 0.5 ml microcentrifuge tube:
   - 25 to 50 ng genomic DNA
   - 200 μM dNTPs
   - 200 pmol primer
   - 5 μl 10 × PCR buffer[a]
   - water to 50 μl
   - 1 unit *Taq* polymerase
2. Spin briefly and overlay with 50 μl mineral oil.
3. Perform 45 cycles as follows: 94°C, 1 min; 35°C, 1 min; 72°C, 2 min.
4. Run 10 μl of the reaction mix through a 1% agarose or Nusieve gel (Section 4.4) in TBE buffer (89 mM Tris, 89 mM boric acid (pH 8.3), 2.5 mM EDTA).

[a] 10 × PCR buffer, see Section 3.3.

Most primers used to amplify birch and *C. fulvum* DNA give bands typically 0.5 to 5 kb in length. More than one band is often observed, with up to three in fungi, and up to ten with higher eukaryotes (J. Williams, pers. comm). A typical result is shown in *Figure 8*. The magnesium concentration is critical (see Section 4.1.3) and varying concentrations sometimes results in an altered profile of amplified bands. It is also important to include controls lacking either target or control DNA.

Our experience concurs with the DuPont group in that the majority of primers which amplify will also reveal polymorphisms between isolates. It is premature to give a precise frequency, but provisionally the chance of discovery of a polymorphism with a primer seems higher than with randomly selected genomic clones and RFLP analysis by hybridization.

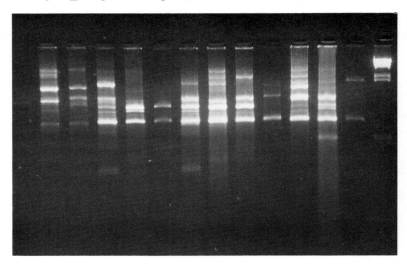

Figure 8. RAPD PCR analysis. Variation in the pattern of bands produced by a single primer and isolates of *C. fulvum* and birch. Lanes 1 and 2, *C. fulvum* race 4 and 5 respectively. Lanes 3 to 12, DNA from individual birch trees including both diploid and tetraploid trees. The weak amplification patterns in lanes 5, 9, and 12 is due to poor quality template DNA. Lane 13, lambda *Hin*d III size markers. Figure produced by Nick Talbot and Diane Howland.

Acknowledgements

Work described in this chapter was funded by SERC, AFRC, and Enichem Americas. S.J.G. holds a Royal Society University Research Fellowsip. We thank Kanjula Senvratne, Adam Corner, Nick Talbot, and Diane Howland for supplying various figures.

References

1. Erlich, H. A. (ed.) (1989). *PCR technology, principles and applications for DNA amplification.* Stockton Press, New York.
2. Innis, M. A., Gelfand, D. H., Sninsky, J. J., and White, T. J. (ed.) (1989). *PCR protocols, a guide to methods and application.* Academic Press, New York.
3. McPherson, M. J., Quirke, P., and Taylor, G. R. (ed.) (1991). *PCR: a practical approach.* IRL, Oxford.
4. Tindall, K .R. and Kunkel, T. A. (1988). *Biochem.*, **27**, 6008.
5. Eckert, K. A. and Kunkel, T. A. (1991). In *PCR: a practical approach* (ed. M. J. McPherson, P. Quirke, and G. R. Taylor), p. 227. IRL, Oxford.
6. Kwok, S. and Higuchi, R. (1989). *Nature*, **339**, 237.
7. Sarkar, G. and Sommers, S. S. (1990). *Nature*, **343**, 27.

Michael J. McPherson, Richard J. Oliver, and Sarah Jane Gurr

8. Marmur, J. (1961). *J. Mol. Biol.*, **3**, 208.
9. McPherson, M. J., Baron, A. J., Jones, K. M., Price, G. J., and Wootton, J. C. (1988). *Protein Engineering*, **2**, 147.
10. Azevedo, M. D., Felipe, M. S., Astolfi Fho., S., and Radford, A. (1991). *J. General Microbiology*, **136**, 2569.
11. Sambrook, J., Fritsch, E. F., and Maniatis, T. (1989). *Molecular cloning: a laboratory manual* (2nd edn.). Cold Spring Harbor Laboratory Press, Cold Spring Harbor, New York.
12. Krug, M. S. and Berger, S. L. (1987). In *Methods in enzymology*, Vol. 152 (ed. S. L. Berger and A. R. Kimmel), p. 316. Academic Press, San Diego.
13. Horton, R. M. and Pease, L. R. (1991). In *Directed mutagenesis: a practical approach* (ed. M. J. McPherson), p. 217. IRL, Oxford.
14. Gelfand, D. H. (1989). In *PCR technology, principles and applications for DNA amplification* (ed. H. A. Erlich), p. 17. Stockton Press, New York.
15. Sarker, G., Kapelner, S., and Sommer, S. S. (1990). *Nucleic Acids Res.*, **18**, 7465.
16. Hung, T., Mak, K., and Fong, K. (1990). *Nucleic Acids Res.*, **18**, 4953.
17. Bachmann, B., Luke, W., and Hunsmann, G. (1990). *Nucleic Acids Res.*, **18**, 1309.
18. Kretz, K. A., Carson, G. S., and O'Brien, J. S. (1989). *Nucleic Acids Res.*, **17**, 5864.
19. Winship, P. R. (1989). *Nucleic Acids Res.*, **17**, 1266.
20. Gyllensten, U. B. and Erlich, H. A. (1988). *Proc. Natl. Acad. Sci. USA*, **85**, 7652.
21. Higuchi, R. G. and Ochman, H. (1989). *Nucleic Acids Res.* **17**, 5865.
22. Stoflet, E. S., Koeberl, D. D., Sarkar, G., and Sommers, S. S. (1988). *Science*, **239**, 491.
23. Crowe, J. S., Cooper, H. J., Smith, M. A., Sims, M. J., Parker, D., and Gewert, D. (1991). *Nucleic Acids Res.*, **19**, 184.
24. Marchuk, D., Drumm, M., Saulino, A., and Collins, F. S. (1991). *Nucleic Acids Res.*, **19**, 1154.
25. Holton, T. A., and Graham, M. W. (1991). *Nucleic Acids Res.*, **19**, 1156.
26. Clackson, T., Gussow, D., and Jones, P. (1991). In *PCR: a practical approach* (ed. M. J. McPherson, P. Quirke, and G. R. Taylor), p. 187. IRL, Oxford.
27. Findlay, J. B. C. and Geissow, M., (ed.) (1989). *Protein sequencing: a practical approach*. IRL, Oxford.
28. Knoth, K., Roberts, S., Poteet, C., and Tamkum, M. (1988). *Nucleic Acids Res.*, **16**, 10932.
29. Sommer, R. and Tautz, D. (1989). *Nucleic Acids Res.*, **17**, 6749.
30. McPherson, M. J., Jones, K. M., and Gurr, S. J. (1991). In *PCR: a practical approach* (ed. M. J. McPherson, P. Quirke, and G. R. Taylor), p. 171. IRL, Oxford.
31. Belyavsky, A., Vinogradova, T., and Rajewsky, K. (1989). *Nucleic Acids Res.*, **17**, 2919.
32. Tam, A. W., Smith, M. M., Fry, K. E., and Larrick, J. W. (1989). *Nucleic Acids Res.*, **17**, 1269.
33. Gurr, S. J. and McPherson, M. J. (1991). In *PCR: a practical approach* (ed. M. J. McPherson, P. Quirke, and G. R. Taylor), p. 147. IRL, Oxford.
34. Lerch, K. (1982). *Journal of Biological Chemistry*, **257**, 6414.

12

Analysis of defence gene transcriptional regulation

MARIA J. HARRISON, ARVIND D. CHOUDHARY, MICHAEL A. LAWTON, CHRISTOPHER J. LAMB, and RICHARD A. DIXON

1. Introduction

A major question in molecular plant pathology is how exposure of plant cells to elicitors, or infection with micro-organisms, induces the appearance of defence gene transcripts leading to the accumulation of antimicrobial barriers. A number of defence response genes have now been cloned (see Appendix A1), and this has facilitated the analysis of transcriptional activation in response to microbial stimuli. Standard techniques of *in vitro* translation, Northern blotting, and RNAase- or S1 nuclease- protection can be used for the measurement of translatable mRNA activities and steady-state mRNA levels in relation to expression of plant defence and are described elsewhere (1, 2). This chapter deals, rather selectively, with direct analysis of transcriptional events and the factors which may interact with the promoters of plant defence genes in order to regulate their expression.

Determination of transcriptional activation requires the analysis of RNA newly-formed in response to the stimuli and is distinct from increased steady state mRNA level, which may be regulated at the level of RNA turnover. Transcriptional activation can be monitored by thiouridine-labelling followed by separation of the newly synthesized RNA on an organomercurial column (3, 4), or, as described here, by measurement of run-off transcription in isolated nuclei (3).

Isolated nuclei are also the source of protein factors which may control defence gene expression. We describe two methods for the isolation of such factors, and outline footprinting techniques which define, *in vitro*, potential binding sites for regulatory proteins. It is important to stress the need to complement *in vitro* analysis of putative transcription factor binding with *in vivo* functional analysis of *cis*-acting sequences. The latter can be performed by studying the expression of deleted or mutated promoter sequences, linked to a suitable reporter gene, either by transient assay in electroporated protoplasts, or after stable transformation in transgenic plants. We outline a system for the transient functional analysis of

147

elicitor-induced promoters in electroporated legume cell protoplasts. Full details of the DNA manipulations and subsequent transformation strategies can be found elsewhere (5–7).

2. Isolation of plant nuclei

A number of techniques are available for the isolation of plant nuclei. *Protocols 1* and *2* describe methods suitable for isolation from cultured cells of bean or alfalfa, or from bean hypocotyl tissue infected with *Colletotrichum lindemuthianum*. These methods are based on the procedures of Willmitzer and Wagner (8) and Luthe and Quatrano (9).

2.1 Nuclei for run-off transcription assays

All manipulations should be carried out at 4°C. *Protocol 1* can be used with frozen plant tissue (which first requires grinding, step 1) or with fresh tissue (start at step 2). A 5–10 fold higher yield of nuclei is expected from fresh cells.

Expected yields of nuclei from bean (*Phaseolus vulgaris*) tissues are in the range of $1-5 \times 10^6$ per g fresh weight of cultured cells, and approximately 10 times less from hypocotyl tissue.

Protocol 1. Isolation of plant nuclei for transcript run-off analysis

1. Grind frozen plant tissue to a fine powder in liquid N_2 with a pestle and mortar.

2. Homogenize tissue (3–10 g fresh weight) in 10 vol. of buffer A[a] for 30 sec with a polytron (Brinkmann, setting 2 or 3).

3. Filter the resultant slurry through three layers of cheesecloth and one layer of nylon mesh (70 μm pore size; Spectromesh).

4. Centrifuge the filtrate 1500 g for 5 min.

5. Gently but thoroughly suspend the pellet in buffer A (5–10 ml), using a plastic Pasteur pipette. Layer on to a discontinuous Percoll gradient[b] and centrifuge at 4000 g for 30 min.

6. With a wide-mouth pipette collect the nuclei, which should band above the 2 M sucrose cushion.

7. Wash the nuclei two times in buffer A, once in buffer A minus spermidine hydrochloride, and once in buffer B[c].

8. Suspend the final nuclear pellet in 0.25 ml buffer B and store in small portions in liquid N_2.

[a] Buffer A contains 0.44 M sucrose, 25 mM Tris–HCl, pH 7.6, 10 mM $MgCl_2$, 10 mM 2-mercaptoethanol, 50 g Dextran T-40 (Pharmacia) per litre, 25 g Ficoll 400 (Pharmacia) per litre, and 2 mM phenylmethylsulphonyl fluoride (PMSF).

Protocol 1. *Continued*

b The gradient (from bottom to top) consists of a 2 M sucrose cushion in gradient buffer (5 ml), then 5 ml layers of 80%, 60%, and 40% (w/v) Percoll in 0.44 M sucrose and gradient buffer. Gradient buffer is 25 mM Tris–HCl, pH 7.6, 10 mM MgCl$_2$.
c Buffer B is 50 mM Tris–HCl, pH 7.8, 5 mM MgCl$_2$, 20 mM 2-mercaptoethanol and 20% (v/v) glycerol.

2.2 Nuclei for isolation of nuclear proteins

Protocol 2 involves release of nuclei after partial protoplasting and has been used successfully with suspension cultured cells of bean and alfalfa for the preparation of nuclear extracts suitable for analysis by gel-retardation and *in vitro* footprinting.

Protocol 2. Isolation of plant nuclei for extraction of nuclear proteins

1. Harvest 7-day-old suspension cultured cells (100–150 g fresh weight) on Miracloth.

2. Transfer to digestion medium,^{a} vacuum infiltrate, and incubate for 30 min in the dark at 28°C on an orbital shaker at 75 r.p.m.

3. Harvest the partially digested cells on a 20 μm nylon mesh, and wash extensively with digestion medium containing no protoplasting enzymes.

4. Carry out all subsequent steps at 0–4°C in solutions supplemented with 100 mM phenylmethylsulphonyl fluoride (PMSF) and 1 μg/ml in each of leupeptin, pepstatin A, chymostatin, and antipain. Wash the cells with 200 ml of ice-cold homogenization buffer.^{b}

5. Thoroughly homogenize the cells in 1–1.5 litres of homogenization buffer supplemented with 0.2% (v/v) Triton X-100, 2% (w/v) Dextran T-40, using a polytron (setting 2 or 3).

6. Pass the slurry through a series of nylon meshes (70 μm over 41 μm over 20 μm) and centrifuge at 1500 g for 5 min.

7. Gently resuspend the pellet in homogenization buffer using a glass rod.

8. Add 1.5 vol of flotation buffer (80% (w/v) Percoll in homogenization buffer, pH 5.8) to the suspension and centrifuge at 2700 g for 15 min. Remove the nuclei from top of the liquid and wash twice in homogenization buffer.

^{a} Digestion medium contains 5 mg/ml Cellulase RS, 0.5 mg/ml Pectolyase Y23, 2 mg/ml bovine serum albumin, 0.7 M mannitol, and 100 mM PMSF in 10 mM MES, pH 5.8.
^{b} Homogenization buffer contains 1.25 M sucrose, 50 mM NaCl, 25 mM EDTA, 0.75 mM spermine HCl, 2.5 mM spermidine phosphate, 100 mM 2-mercaptoethanol, and 0.04% (v/v) Triton X-100 in 50 mM MES, pH 5.2.

3. Nuclear transcript run-off analysis

Detailed protocols for the analysis of transcription in isolated mammalian cell nuclei have been described elsewhere (2). *Protocol 3* has been used successfully for analysis of defence gene transcription in isolated bean nuclei.

Protocol 3. Transcript run-off analysis with plant nuclei

1. Add nuclei (up to 2×10^7; *Protocol 1*) to a final volume of 200 μl containing 20 mM Tris–HCl, pH 7.8, 75 mM $(NH_4)_2SO_4$, 5 mM $MgCl_2$, 2 mM $MnCl_2$, 0.5 mM each of ATP, CTP, and GTP, 1 mM dithiothreitol, 8% (v/v) glycerol, 1 mCi/ml $[\alpha-^{32}P]UTP$ (600 Ci/mmol, New England Nuclear Corp.) and 750 U/ml RNAasin (Promega Biotech).

2. Incubate for 30 min at 26°C.

3. Terminate the reaction by addition of 50 μg of tRNA (*E. coli*, RNAase-free; Boehringer Mannheim) and 2 μg of RQ1 DNAase (Promega Biotech).

4. Remove a 5 μl aliquot, and spot on to a Whatman 3MM paper strip for measurement of total incorporation of UTP into acid-precipitable material. Wash strips for 10 min in an ice-cold solution containing 10% (w/v) trichloroacetic acid (TCA) and 1% (w/v) $Na_4P_2O_7$. Wash twice in an ice-cold solution containing 5% (w/v) TCA and 1% (w/v) $Na_4P_2O_7$, and finally rinse in absolute ethanol. Dry and determine radioactivity by liquid scintillation counting.

5. Isolate RNA from the remainder of the assay mixture as described in ref. 10, and precipitate with ethanol.

6. Dissolve the ethanol precipitate in hybridization buffer[a], and denature at 80°C for 5 min prior to hybridization to immobilized defence gene cDNA sequences.

[a] Hybridization buffer is: 50% (v/v) formamide, 4 × SSPE, 1 × Denhardt solution, 0.2% (w/v) SDS, 0.1 μg/ml tRNA, and 0.1 μg/ml polyadenylic acid (Calbiochem). 1 × SSPE is 0.18 M NaCl, 10 mM NaPO$_4$, pH 7.7, 1 mM EDTA. 1 × Denhardt solution is 0.02% (w/v) bovine serum albumin, 0.02% (w/v) Ficoll, 0.02% (w/v) polyvinylpyrrollidone.

---

4. Isolation and assay of *trans*-acting factors

4.1 Introduction

Several techniques are now available for measuring binding of plant factors to specific DNA sequences in the regulatory regions of inducible genes. These involve the incubation of isolated plant nuclear proteins with a labelled probe containing the putative target sequence for the binding factor. Analysis *in vitro*, by gel retardation or footprinting, is relatively straightforward technically, but requires careful attention to controls. Such methods should always be complemented by functional promoter analysis *in vivo*. In some cases, levels of DNA-binding proteins may be induced by applied hormonal or environmental factors (11), or may show cell-type specificity (12). The establishment of such correlations, based purely on *in vitro* binding studies, may not always indicate that the DNA sequence to which binding is being studied is functionally active in the manner predicted.

The best controls for gel retardation assays are mutant oligonucleotides where only small changes have been made to the DNA binding sequence (13). Such controls are most meaningful if it can be shown that the introduced mutation affects the activity of the promoter *in vivo*. We have found that use of less stringent DNA competitors can lead to positive identification of specific binding which later proves artefactual (14).

4.2 Preparation of nuclear extracts

Protocol 4 describes the extraction of proteins from nuclei prepared from partially protoplasted cells by the method in *Protocol 2*. As different nuclear proteins are solubilized as a function of salt concentration, it is necessary to test a range of salt concentrations in order to optimize extraction conditions. Successful extraction of the required *trans*-acting factor is assessed by the gel retardation assay as described in Section 4.3.

Protocol 4. Extraction of plant nuclear proteins

1. Prepare lysis buffers with increasing concentrations of sodium chloride as follows; 20 mM HEPES pH 7.9, 25% (v/v) glycerol, 1.5 mM $MgCl_2$, 0.2 mM EDTA, 0.1 M to 0.5 M NaCl in 0.1 M steps.
   Immediately prior to use add: PMSF to 0.5 mM, DTT to 0.5 mM, proteinase inhibitors leupeptin, chymostatin, pepstatinA, antipain, each to 0.5 μg/ml.

2. Add an approximately equal vol. of 0.1 M NaCl lysis buffer to the pelleted nuclei from *Protocol 2* and resuspend vigorously. If necessary lysis may be assisted by 5–10 strokes in a Dounce homogenizer.

3. Incubate on ice for 10–20 min.

4. Pellet the lysed nuclei by centrifugation e.g. at 25 000 r.p.m. (Beckman ultracentrifuge Ti 50 rotor) or equivalent for 30 min.

5. Remove the supernatant and dialyse for 6 h against Dignam's dialysis buffer.[a]

6. Resuspend the pellet in 0.2 M NaCl lysis buffer and repeat the incubation and centrifugation procedure. Continue the above sequence up to 0.5 M NaCl lysis buffer.

Protocol 4. *Continued*

7. Store the dialysed supernatants at −80°C in small aliquots and assay for binding activity using gel retardation assays (see *Protocol 5*).

a Dignam's dialysis buffer is 20 mM HEPES, pH 7.9, 20% (v/v) glycerol, 0.1 M KCl, 0.2 mM EDTA, 0.5 mM PMSF, 0.5 mM DTT.

4.3 Gel retardation assays

The gel retardation assay (GRA) (also known as gel binding or mobility shift assay) is a rapid and sensitive technique for the detection of proteins that are capable of binding to a given DNA sequence. The assay can be performed with crude or purified nuclear extracts. A labelled DNA fragment containing the binding site sequence is incubated with nuclear protein and the mixture is separated by polyacrylamide gel electrophoresis. The free DNA probe migrates rapidly towards the bottom of the gel while probe with protein bound to it migrates more slowly. The gel is dried and the relative positions of the DNA and DNA–protein complexes are visualized by autoradiography. A variety of binding buffers and polyacrylamide gel systems have been used to analyse binding interactions in a wide range of *trans*-acting factor–promoter systems. *Protocol 5* has been used to identify nuclear factors binding to plant defence gene promoter sequences.

Protocol 5. Assay of DNA–protein interactions by gel retardation

A. *Binding reaction*

1. Mix the following components in a microcentrifuge tube.
   - 1 μl 10 × MM buffer*a*
   - 0.1–1 ng ^{32}P-labelled DNA probe (10^3–10^4 d.p.m.)
   - 2 μl poly (dI.dC) (dI.dC)
   - up to 10 μg of nuclear protein
   - distilled H_2O to a final volume of 10 μl

   The nuclear protein should be added last to ensure competition for binding between probe and poly (dI.dC) (dI.dC).

2. Incubate on ice for 30 min.

B. *Gel electrophoresis*

3. Add 1 μl of loading dye to each sample and electrophorese on a low ionic strength (0.5 × TBE), 4% polyacrylamide gel (30 : 1) at 5–10 volts/cm. Dry the gel and expose to X-ray film.

a 10 × MM buffer contains 100 mM Tris–HCl, pH 7.5, 500 mM NaCl, 10 mM EDTA, 40% (v/v) glycerol, 10 mM dithiothreitol (DTT), and 5 mM PMSF. Add DTT and PMSF immediately prior to use.

The amount of poly (dI.dC) (dI.dC) required depends on the amount of protein in the reaction and must be estimated for each protein and probe preparation by titration. The range is usually $0.1–10$ μg per assay. If the concentration is too low the probe will be retarded non-specifically to the top of the gel. Sometimes non-specific retardation looks like a background smear. Too much poly (dI.dC) (dI.dC) may result in lowered or abolished retardation of the probe. Other non-specific competitor DNAs such as sonicated calf thymus or salmon sperm DNA may also be used.

The ionic strength of the gel and buffering system can affect the result. High ionic strength gels and buffer may prevent the detection of weak protein–DNA interactions; however, better resolution of strong protein–DNA interactions is often obtained on a higher ionic strength system. The amount of cross-linking in the gel may also be altered, e.g. to $60:1$ or $80:1$, to distinguish between closely migrating species.

4.3.1 Competition gel retardation assays

Competition gel retardation assays are used to demonstrate the specificity of the binding interactions. An excess of unlabelled specific competitor DNA containing the same binding site sequence is included in the reaction. The amount of specific competitor required to prevent formation of the labelled DNA–protein complex is estimated by titration. A molar excess of $50–100$ of unlabelled specific DNA over labelled probe is usually sufficient. An equivalent amount of unlabelled non-specific competitor should not affect the formation of a specific complex (*Figure 1*). The competition assay can be modified to determine the 'off' rate (for complex dissociation) by pre-binding to labelled probe and measuring the changes in the ratio of bound/free fragment following the addition of cold competitor.

4.4 DNAase I footprinting

In *in vitro* footprinting a labelled DNA fragment containing a putative protein binding site is incubated with a nuclear protein preparation. Regions to which protein are bound protect the DNA from attack by DNAase I. After digestion with DNAase I, samples are analysed on a sequencing gel and the protected region or regions appear as footprints on a ladder resulting from the random cutting of the DNA at each unprotected position. A normal sequencing ladder generated from the DNA fragment under investigation is run in parallel to enable the protected regions to be mapped to the sequence. The technique is described in *Protocol 6* and further details of the method are given in ref. 15. In this method, unlike the GRA, bound and free forms are not resolved and each footprint track represents the aggregate pattern of all forms of the fragment. In our experience, footprints are clearly detectable when $> 80\%$ of the fragment exists in the bound form (as judged by GRA). If there is an insufficient amount of the bound form, the

A B C D E

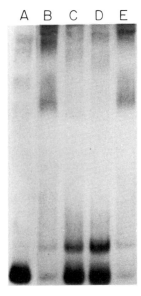

Figure 1. Competition gel retardation. A ^{32}P-labelled probe comprising a 200 bp fragment from the promoter region of a bean chalcone synthase gene (track A) was incubated with nuclear extract from bean leaves (tracks B–E) in the presence of 2.5 µg poly (dl.dC) (dl.dC) per 10 µl assay. Data show the effects of no competitor (B), 50 molar excess unlabelled 200 bp fragment (C), 50 molar excess of another unlabelled fragment from the CHS promoter containing a protein-binding sequence common to the probe fragment (D), and 50 molar excess non-specific CHS cDNA sequence (E).

bound and free forms are isolated from an acrylamide gel and treated separately with DNAase I.

Protocol 6. DNAase I footprinting

1. Prepare a radioactively end-labelled DNA probe (approximately 0.1 ng DNA, 50 c.p.s. ^{32}P per reaction).

2. Mix the following components in two microcentrifuge tubes and incubate on ice for 25 min.

Control reaction	*Footprinting reaction*
6.5 µl H$_2$O	6.5 µl H$_2$O
25 µl master buffer[a]	20 µl master buffer
5 µl supplementary buffer[b]	5 µl supplementary buffer
2.5 µl poly (dI.dC) (dI.dC) (2.5 µg/µl stock)	2.5 µl poly (dI.dC) (dI.dC)

Protocol 6. *Continued*

10 μl 10% (w/v) polyvinyl alcohol (PVA)	10 μl 10% (w/v) PVA
1 μl end-labelled probe	1 μl end-labelled probe
	5 μl nuclear extract ~1 μg protein/μl)

The ratio of nuclear extract, probe, and poly (dI.dC) (dI.dC) must be determined for each individual case.

3. Prepare dilutions of DNAase I in dilution buffer[c] (usually between 7.5 and 15 μg/ml, but concentrations must be optimized for each individual case. The control reaction which does not contain nuclear extract may require a slightly lower dilution of DNAase I to achieve the equivalent amount of digestion to the reactions containing nuclear extract.) Immediately prior to use mix equal volumes of diluted DNAase I and 100 mM $MgCl_2$.

4. Add 5 μl of DNAase I/$MgCl_2$ to each reaction and incubate at 25°C for 1 min.

5. Add 50 μl of stop solution[d] and 100 μl phenol. Vortex briefly.

6. Phenol extract the reaction mixtures twice and chloroform extract once.

7. Ethanol precipitate the DNA by adjusting the solution to 0.3 M sodium acetate and add 2 vols ethanol. Incubate at −20°C for 15 min.

8. Collect the precipitated DNA by centrifugation, wash with 70% ethanol, and dry the pellet under vacuum.

9. Resuspend the pellet in a standard DNA sequencing loading buffer. Heat at 100°C for 5 min. Cool on ice. Analyse by electrophoresis on a standard 6–8% sequencing gel along with a normal sequence ladder of the DNA being footprinted.

[a] Master buffer contains 994.5 μl Dignam's dialysis buffer (minus PMSF and DTT; see *Protocol 4*), 5 μl 0.1 M PMSF and 0.5 μl 1 M DTT.
[b] Supplementary buffer contains 82 μl H_2O, 15 μl 1 M HEPES, pH 7.9, 2.5 μl 0.1 M PMSF, and 0.5 μl 1 M DTT.
[c] Dilution buffer contains 10 mM HEPES, pH 7.9, 150 mM NaCl, 5 mM $MgCl_2$, 5 mM $CaCl_2$, and 1 mg/ml BSA.
[d] Stop solution contains 20 mM EDTA, 1% (w/v) SDS, 0.2M NaCl, 250 μg/ml tRNA.

4.5 Chemical footprinting

Footprints obtained using DNAase I may extend slightly beyond the contact sites between a *trans*-acting factor and its DNA binding site due to the size of the DNAase I molecule itself. Furthermore, the enzyme does not always cut DNA in a completely sequence-independent manner. In order to circumvent these problems and to increase resolution a number of techniques have been developed in which the phosphodiester bonds of the DNA are cleaved by chemical rather than enzymatic reactions. These include hydroxyl radical footprinting (16), copper phenanthroline footprinting (17), and methidium propyl EDTA-iron(II) footprinting (18).

4.6 Strategies for the cloning of plant *trans*-acting factors

The first reports of the cloning of genes for mammalian transcription factors used standard strategies similar to those used for the cloning of most other types of genes, namely the purification of the protein and the use of amino acid sequence data from either the N-terminus or from internal peptides for the design of oligonucleotides for screening cDNA or genomic libraries. This approach has now been successfully used to clone many transcription factor genes from mammals and yeast (19–22). However, to date, there have been no documented reports of the purification and protein-sequencing of transcription factors from plants.

A second approach which has proved successful for isolating cDNA clones encoding transcription factors was originally developed by Singh *et al.* (23). This strategy, derived from λgt11/antibody screening protocols, involves screening expression libraries using the DNA recognition sequence of the binding protein as a probe (24). Use of this approach has resulted in the isolation of the first clones for plant transcription factors; a tobacco protein involved in the regulation of transcription from the cauliflower mosaic virus 35S promoter (25) and a binding protein from wheat which interacts with a histone gene promoter (26).

5. *In vivo* analysis of *cis*-acting sequences and *trans*-acting factors

5.1 Introduction

In the absence of an *in vitro* transcription system for plant cells, functional analysis of putative *cis*-acting regulatory sequences in plant genes must be performed in living cells. The DNA sequence of interest (intact, deleted, or mutated promoter elements linked to a suitable reporter gene) is introduced into plant tissues by a variety of methods, such as stable transformation mediated by *Agrobacterium tumefaciens*, or electroporation of protoplasts. Detailed descriptions of vectors, reporter genes, and transformation protocols are given elsewhere (5, 7, 27).

5.2 Transient promoter expression assays

Transient gene expression assays have the advantage over analysis by stable transformation of greater rapidity. However, it must be realized that the behaviour of a gene promoter in electroporated protoplasts can sometimes differ from the cell-type specific expression observed in cells of the intact plants.

Furthermore, when studying expression of defence response genes, the preparation of protoplasts by enzymic digestion may result in the elicitation of the protoplasts as a result of the release of cell wall components. This effect varies with species and the conditions used, as for example, with parsley protoplasts in which no elicitation occurs (28), soybean protoplasts which are fully elicited, and therefore unresponsive to added elicitors (29), and soybean protoplasts which are elicited but where the response returns to basal level within a few hours, and the protoplasts are then responsive to added elicitors (30). *Protocol 7* describes a transient assay system used to study defence gene promoters in electroporated alfalfa protoplasts (31). The basal level of elicitation and subsequent responsiveness to added elicitor vary with the age of the suspension culture from which the protoplasts are derived. Suspensions up to 9 passages after initiation from callus yield protoplasts with low basal expression of an introduced chimeric bean chalcone synthase promoter/reporter construct, and are highly responsive to added elicitors (*Figure 3*). This system allows functional analysis of elicitor-responsive elements by comparison of the expression of deleted or mutated promoter sequences.

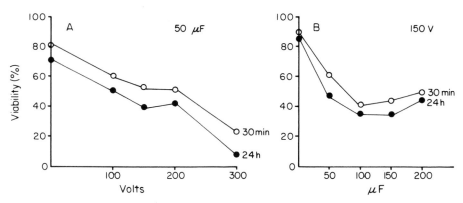

Figure 2. Viability of alfalfa protoplasts after electroporation using a BTX Transfector 300 unit. Samples (500 μl) containing 10^7 protoplasts were electroporated according to the details in *Protocol 7*; the data show the effects of varying electroporation voltage at constant capacitance (A), or altering capacitance at constant voltage (B). Viability was assessed by Evans blue staining at 30 min and 24 h post-electroporation. Conditions of 50 μF and 150 V give the best uptake/expression of DNA in alfalfa protoplasts under the experimental conditions described.

Bacterial chloramphenicol acetyltransferase (CAT) is the most commonly used reporter gene for transient expression assays. The chimaeric genes for electroporation would therefore consist of defence gene promoter–CAT–nopaline synthase 3′ terminator constructs in a suitable plasmid such as pUC19. A new, rapid assay for CAT activity has recently been described (32).

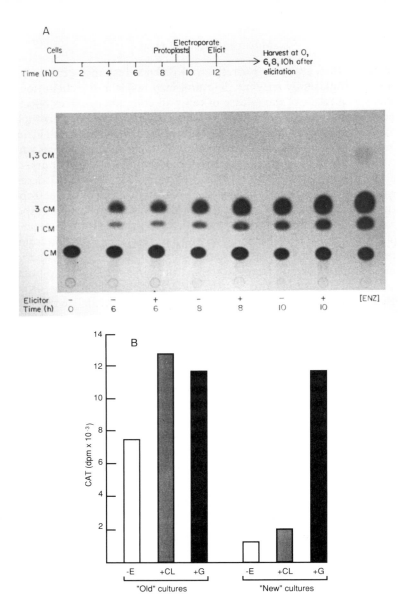

Figure 3. (A) Expression of chloramphenicol acetyltransferase (CAT) activity in alfalfa protoplasts prepared, electroporated and exposed to a fungal elicitor preparation according to the timing protocol shown. The electroporated gene construct consisted of a bean chalcone synthase promoter–CAT–nopaline synthase 3′ terminator fusion (30, 31). Note the high level of expression of CAT in unelicited protoplasts. The effect of fungal elicitor is to increase expression by approximately 40% in these protoplasts obtained from cell suspension cultures which had been through over 10 passages after initiation from callus. By comparison (B), protoplasts from newly initiated cultures have low basal CAT expression, a similar relative response to fungal elicitor but a far greater responsiveness to the abiotic reduced glutathione. −E=no elicitor, +CL=plus cell wall elicitor from *Colletotrichum lindemuthiamum* (50 μg glucose equivalents per ml), +G=plus glutathione (1 mM final concentration).

Protocol 7. Functional analysis of a defence gene promoter in
electroporated protoplasts

The yield, viability, and responsiveness of protoplasts may be cultivar-
dependent. We have found cultivar Calwest 475 to be satisfactory.

1. Initiate suspension cultures of alfalfa (*Medicago sativa* L.) from callus
   cultures in modified SH medium (33) supplemented by 1.8 mg/l
   p-chlorophenoxyacetic acid, 0.5 mg/litre 2,4-D, 0.5 mg/l kinetin, 100 mg/l
   L-serine, 800 mg/l, L-glutamine, and 1 mg/l adenine. Grow in the dark at
   25°C with constant shaking and subculture (10 ml culture into 40 ml fresh
   medium) every 7 days. (Ref. 33 contains full details on the initiation and
   maintenance of plant callus and cell suspension cultures.)

2. Prepare and filter-sterilize the protoplasting enzyme solution which
   contains 1% (w/v) driselase (Sigma), 1% (w/v) cellulase (Onozuka Rs),
   0.5% (w/v) macerozyme (Onozuka R10), 0.5% (w/v) hemicellulase (Sigma),
   and 0.4 M mannitol in suspension culture medium pH 5.8.

3. Incubate cultured cells (1 g, 5–6 days after subculture) with 10 ml protoplasting enzyme solution in 100×15 mm Petri dishes with shaking
   (40 r.p.m.) at 25°C in the dark for 12–14 h.

4. Separate protoplasts from undigested cells and debris by successive passages
   through 70, 40, and 30 μm nylon mesh filters. Collect protoplasts by
   centrifugation at 100 g for 10 min, and wash three times with W5 medium,
   pH 5.8.[a]

5. Count the protoplasts in a haemocytometer and determine viability using
   Evans blue stain (34). The yield should be approximately 1×10^6 per g fresh
   weight, with viability in excess of 90% and no cell wall residues should be
   present as observed by Calcofluor white staining (35). If these criteria are not
   met the protoplasts should not be used.

6. Re-suspend protoplasts at a density of 2×10^7 per ml in MsMg solution.[b]

7. Heat-shock the protoplasts for 5 min at 45°C and then bring to room
   temperature by incubating on ice for 2 min.

8. Transfer 500 μl aliquots of heat-shocked protoplasts to 1 ml plastic
   spectrophotometer cuvettes containing the DNA to be electroporated
   (10–60 μg plasmid containing the defence gene promoter construct plus
   50 μg carrier calf thymus DNA). Incubate at room temperature for 10 min.

9. Add 200 μl of 40% PEG solution[c] and incubate on ice for 10 min.

10. Electroporate using a commercially available electroporation apparatus
    such as a BTX Transfector 300 unit (BTX, San Diego).

11. Incubate protoplasts on ice for 10 min then at room temperature for 10 min
    before diluting with 5 vol. 0.2 M $CaCl_2$ 0.4 M mannitol. Centrifuge for
    5 min at 50 g, wash with protoplast culture medium (36) and culture in 2 ml
    of protoplast culture medium in the dark at 25°C.

Protocol 7. *Continued*

12. Protoplasts may be treated with elicitors at various times up to 20 h after preparation and subsequently harvested for assay of CAT activity.

---

[a] W5 medium is 154 mM NaCl, 125 mM CaCl$_2$, 5 mM KCl, 5 mM glucose, 0.4 M mannitol, pH 5.8–6.0.
[b] MsMg solution is Murashige and Skoog salts (37), 0.4 M mannitol, 30 mM/MgCl$_2$, 0.1% (w/v) MES, pH 5.8.
[c] 40% PEG solution is 40% (v/v) polyethylene glycol (mol. wt 8000) 0.1% (w/v) MES, 0.4 M mannitol, 30 mM MgCl$_2$, pH 7.0.

Electroporation conditions must be determined experimentally. For alfalfa protoplasts using the BTX apparatus the pulse is delivered from capacitors of 50 μF charged to 150 V. A useful guide is that electroporation conditions which result in an approximately 30–50% decrease in protoplast viability will probably result in suitable uptake of DNA. *Figure 2* demonstrates the relationship between electroporation conditions and viability for alfalfa protoplasts.

5.3 *In vivo* genomic footprinting

In vivo genomic footprinting is a powerful but somewhat difficult technique for observing the binding of nuclear proteins to *cis*-acting elements in the living cell. It provides excellent confirmation for the results of *in vitro* binding and functional assays. Alternatively, it may be used as a first step in defining putative *cis*-acting elements prior to functional analysis. The basis of the method is the protection of DNA from reaction with dimethyl sulphate *in vivo* when binding proteins are attached. Total genomic DNA is then fragmented, extracted, cleaved at the modified bases, fractionated by electrophoresis, transferred to a hybridization membrane, and hybridized to a labelled probe complementary to the region of DNA under investigation. The reader is referred elsewhere for details of the basic method (38–40), its application to the study of regulatory elements in the light-regulated parsley chalcone synthase (CHS) promoter, and the elicitor/light-induced parsley phenylalanine ammonia-lyase (PAL) promoter (41, 42), and recent modifications, using the polymerase chain reaction, to increase the sensitivity of the method (43).

References

1. Hames, B. D. and Higgins, S. J. (ed.) (1985). *Nucleic acid hybridisation, a practical approach.* IRL, Oxford.
2. Hames, B. D. and Higgins, S. J. (ed.) (1984). *Transcription and translation, a practical approach.* IRL, Oxford.

3. Marzluff, W. F. and Huang, R. C. C. (1984). In *Transcription and translation, a practical approach* (ed. B. D. Hames and S. J. Higgins), p. 89. IRL, Oxford.
4. Cramer, C. L., Ryder, T. B., Bell, J. N., and Lamb, C. J. (1985). *Science*, **227**, 1240.
5. Draper, J., Scott, R., Armitage, P., and Walden, R. (1988). *Plant genetic transformation and gene expression. A laboratory manual.* Blackwell, Oxford.
6. Sambrook, J., Fritsch, E. F., and Maniatis, T. (1989). *Molecular cloning—a laboratory manual* (2nd edn). Cold Spring Harbor Laboratory Press.
7. Glover, D. M. (ed.) (1985). *DNA cloning, a practical approach*, Vol. 2. IRL, Oxford.
8. Luthe, D. S. and Quatrano, R. S. (1980). *Plant Physiol.*, **65**, 305.
9. Willmitzer, L. and Wagner, K. G. (1981). *Exp. Cell Res.*, **135**, 69.
10. McKnight, G. S. and Palmiter, R. D. (1979). *J. Biol. Chem.*, **254**, 9050.
11. Holdsworth, M. J. and Laties, G. G. (1989). *Planta*, **179**, 17.
12. Jofuku, K. D., Okamuro, J. K., and Goldberg, R. B. (1987). *Nature*, **328**, 734.
13. Kuhlemeier, C., Cuozzo, M., Green, D. J., Goyvaerts, E., Ward, K., and Chua, N-H. (1988). *Proc. Natl. Acad. Sci. USA*, **85**, 4662.
14. Dixon, R. A., Harrison, M. J., Jenkins, S. M., Lamb, C. J., Lawton, M. A., and Yu, L. (1990). In *Plant gene transfer* (ed. C. J. Lamb and R. Beachy), p. 101. Alan R. Liss, New York.
15. Jackson, P. D. and Gelsenfeld, G. (1985). *Proc. Natl. Acad. Sci. USA*, **82**, 2296.
16. Landolfi, N. F., Ming Yin, X., Capra, J. D., and Tucker, P. W. (1989). *Biotechniques*, **7**, 500.
17. Tullius, T. D. and Dombraski, B. A. (1986). *Proc. Natl. Acad. Sci. USA*, **83**, 5469.
18. Kuwabara, M. D. and Sigman, D. S. (1987). *Biochemistry*, **26**, 7234.
19. Briggs, M. R., Kadonaga, J. T., Bell, S. P., and Tjian, R. (1986). *Science*, **234**, 47.
20. Wiliams, T., Adman, A., Luscher, B., and Tjian, R. (1988). *Genes and Development*, **2**, 1557.
21. Santoro, C., Mermod, N., Andrews, P. C., and Tjian, R. (1988). *Nature*, **334**, 535.
22. Frain, M., Swart, G., Monaci, P., Nicorai, A., Stampfli, S., Frank, R., and Cortese, R. (1989). *Cell*, **59**, 145.
23. Singh, H., LeBowitz, J. H., Baldwin, Jr., A. S., and Sharp, P. A. (1988). *Cell*, **52**, 415.
24. Singh, H., Clerc, R. G., and LeBowitz, J. H. (1989). *Biotechniques*, **7**, 252.
25. Katagiri, F., Lam, E., and Chua, N-H. (1989). *Nature*, **340**, 727.
26. Tabata, T., Takase, H., Takayama, S., Mikami, K., Nakatzuka, A., Kawata, T., Nakayama, T., and Iwabuchi, M. (1989). *Science*, **245**, 965.
27. Jefferson, R. A., Kavanagh, T. A., and Bevan, M. W. (1987). *EMBO J.*, **6**, 3901.
28. Dangl, J. L., Hauffe, K. D., Lipphardt, S., Hahlbrock, K., and Scheel, D. (1987). *EMBO J.*, **6**, 2551.
29. Mieth, H., Speth, V., and Ebel, J. (1986). *Z. Naturforsch*, **41c**, 193.
30. Dron, M., Clouse, S. D., Lawton, M. A., Dixon, R. A., and Lamb, C. J. (1988). *Proc. Natl. Acad. Sci. USA*, **85**, 6738.
31. Choudhary, A. D., Kessmann, H., Lamb, C. J. and Dixon, R. A. (1990). *Plant Cell Reports*, **9**, 42.
32. Seed, B., Sheen, J. Y. (1988). *Gene*, **67**, 271.
33. Dixon, R. A. (ed.) (1985). *Plant cell culture, a practical approach.* IRL, Oxford.
34. Graff, D. F. and O'Kong-O-Ogola, O., (1971). *J. Exp. Bot.*, **22**, 756.
35. Nagata, T. and Takebe, I. (1970). *Planta.*, **92**, 12.
36. Menczel, L., Nagy, F., Kiss, Z., and Maliga, P. (1981). *Theor. Appl. Genet.*, **59**, 191.
37. Murashige, T. and Skoog, F. (1962). *Physiol. Plant.*, **15**, 473.

38. Neilsen, P. E. (1989). *Bioessays*, **11**, 152.
39. Church, G. M. and Gilbert, W. (1984). *Proc. Natl. Acad. Sci. USA*, **81**, 1990.
40. Saluz, M. P. and Jost, J. P. (1989). *Anal. Biochem.* **176**, 201.
41. Schulze-Lefert, P., Dangl, J. L., Becker-Andre, M., Hahlbrock, K., and Schulz, W. (1989). *EMBO J.*, **8**, 651.
42. Lois, R., Dietrich, A., Hahlbrock, K., and Schulz, W. (1989). *EMBO J.*, **8**, 1641.
43. Salva, M. P. and Jost, J. P. (1989). *Nature*, **338**, 277.

13

In situ hybridization in plants

DAVID JACKSON

1. Introduction

In situ hybridization techniques provide an exquisitely sensitive method to study tissue-specific or organ-specific patterns of gene expression.

The techniques described in this chapter and based on refs 1–4, have been developed to study the expression of several genes involved in the biosynthesis of pigment (anthocyanin) in different floral organs (petal, style, stamen) of *Antirrhinum majus*. The methods involve hybridization of tritium-labelled single stranded RNA probes to mRNA in sections of fixed and wax embedded plant material. Hybridization is visualized using photographic emulsion, where radioactive emissions produce latent images in the emulsion which form silver grains when developed. The silver grains are visualized by dark-field microscopy.

The techniques are applicable to a wide range of plant tissue and to different probes, including those involved in plant–pathogen interactions. For example, these techniques have been used at the John Innes Institute to study:

- storage protein gene expression in developing pea embryos (A. Hauxwell, personal communication);

- tobacco rattle virus RNA in and around infection sites on tobacco leaves (V. Ziegler-Graff and D. Baulcombe, unpublished results);

- polygalacturonase mRNA in tomato pericarp tissue (C. Watson, unpublished results);

- genes expressed specifically in developing barley embryos (L. Smith, D. Bowles, unpublished results);

- chalcone synthase mRNA around necrotic lesions on leaves of *A. majus* (D.J., unpublished results).

The techniques described in this chapter have been divided into sections (fixation, pretreatments, etc.) for clarity, but many steps are closely interrelated. Alteration of a step in one section can have profound effects on steps in other sections. Standard practices for working with RNA should be employed at all stages.

2. Fixation and wax embedding

Fixation is a critical step as the RNA is held in a 3-dimensional array, attached to proteins and other macromolecules. Poorly fixed material will give little or no *in situ* signal even with probes for highly abundant mRNAs. Fixation should provide RNA retention (i.e., give sufficient hold to the RNA such that it is not washed out of the sections in later steps) yet leave it accessible to the probe. In practice, the sections of fixed material undergo a limited protease digestion to increase probe accessibility, and the probe is hydrolysed into small fragments which can penetrate the section more easily.

There are two commonly used types of fixatives, those that precipitate non-specifically and the aldehydes. Aldehyde fixatives give superior RNA retention and accessibility to the probe (5). Formaldehyde and glutaraldehyde have been used as fixatives for *in situ* hybridization, either individually or as mixtures. I find little difference in the *in situ* signal obtained using either of these but prefer to use formaldehyde fixed material because anthocyanin pigment is retained and this makes the plant tissues more visible during sectioning. With glutaraldehyde fixed material, the pigment is washed out during the ethanol dehydration steps.

The most important aspect of fixation is the penetration of fixative into the material. Most plant organs are surrounded by a waxy cuticle and float on the surface of aqueous fix solutions. To overcome this, the fix can be infiltrated under vacuum. The plant material is held under the surface of the fix (by using wire gauze with a small weight on top) and a vacuum is then applied and released slowly (to avoid tissue damage). I use a 2-stage rotary pump connected to a dessicator operating at around 40 Pascals, and infiltrate the material until it sinks when the vacuum is released (usually 1–5 min). Formaldehyde is volatile, so the fix solution should be changed after vacuum infiltration. Material should be dissected into small pieces immediately before placing into the fix. It is advisable to try a range of sizes; as a general guide leaf pieces should not exceed about 5 mm × 10 mm and more solid structures should be cut so that at least one dimension is less than 1 mm.

After fixation, the material is dehydrated through an ethanol series. The ethanol solutions can also be vacuum infiltrated.

The dehydration, clearing, and wax embedding steps include gradual changes to avoid tissue damage due to excessive shrinkage or swelling.

Protocol 1. Fixation and wax embedding

Day 1

1. Prepare 4% formaldehyde fix solution just before use: take 100 ml phosphate buffered saline (PBS) (130 mM NaCl, 7 mM Na_2HPO_4, 3 mM NaH_2PO_4) and adjust to pH 11 using NaOH. Heat to 60°C then add 4 g paraformaldehyde and stir until dissolved (1–2 min). Cool on ice and then readjust to pH 7.0 using H_2SO_4.

David Jackson

Protocol 1. *Continued*

Caution

- formaldehyde vapour is toxic—use a fume hood
- Do not use HCl to adjust the pH as the combination of HCl and formaldehyde releases a powerful carcinogen

2. Cut plant material and place in fix on ice.[a] Vacuum infiltrate then renew fix (see step 1). Leave at 4°C overnight.

Day 2

3. Pour off fix and replace with ice cold 0.85% NaCl. Leave on ice for 30 min, swirling occasionally.

4. Repeat for 90 min each with 50%, 70%, 85% ethanol, each containing 0.85% (w/v) NaCl, then 95% ethanol (no NaCl) and 100% ethanol.[b]

5. Renew 100% ethanol and leave overnight at 4°C.

Day 3

6. Renew 100% ethanol and leave for 2 h at room temperature. Replace with 50% ethanol:50% Histoclear[c] (Agar Aids Ltd) and leave for 1 h.

7. Repeat three times with 100% Histoclear. Add Paraplast chips (BDH) to approximately half the vol. of the Histoclear. Leave overnight.

Day 4

8. Incubate at 40–50°C until Paraplast chips dissolve (this takes a few hours). Replace with molten Paraplast at 60°C.[d] Leave overnight.

Days 5–7

9. Renew Paraplast each morning and evening. On day 7 make wax blocks by pouring some Paraplast with the plant material into a mould then float the mould on water to solidify the wax. Store wax blocks at 4°C. They are stable for several months.

[a] Use a large excess of each solution (e.g. for ten 5 × 10 mm leaf pieces use 25 ml).
[b] 100% ethanol is stored over molecular sieves.
[c] Histoclear is a less-toxic alternative to xylene.
[d] Paraplast should be freshly melted before use and should not be heated to temperatures above 60°C as this destroys synthetic polymers which are added to Paraplast to aid sectioning.

Protocol 2. Preparation of slides and coverslips

1. Soak slides in concentrated nitric acid for 30 min. Use 'frosted end' slides, these can be permanently marked using pencil.

2. Wash for at least 1 h in several changes of distilled water, drain.

3. Wash for 15 min in acetone.

4. Drain, then bake at 180°C for at least 2 h.

165

Protocol 2. *Continued*

5. When the slide is cool, add 5 μl poly-L-lysine hydrobromide (mol. wt 150–300 000) in water and draw into a film over the slides using a coverslip.

6. Leave slides on a 42°C hotplate overnight to dry, then store in a box with dessicant at 4°C.

7. To prepare coverslips, wash in acetone and bake as in step 4.

Protocol 3. Sectioning of wax embedded material

Good sectioning requires considerable practise (and patience!). It is a good idea to start by spending a few hours with someone who is skilled in wax sectioning.

1. Cut the wax block to a trapezoid shape, leaving about 2 mm of wax around the plant material.

2. Mount the block such that the longer of the two parallel faces is at the bottom (it is the first to strike the blade).

3. Cut ribbons of sections of 10 μm thickness. Float a ribbon on drops of sterile water on a poly-L-lysine coated slide. Place slide on a 42°C hotplate for a few minutes until the ribbon flattens out.

4. Drain off excess water and leave on the hotplate overnight to dry. The slides can now be stored in a box with dessicant at 4°C and are stable for at least 2 months.

---

3. Prehybridization

The steps described in *Protocol 4* are designed (a) to dewax and rehydrate the sections, (b) to permeabilize the tissues, thus enhancing probe penetration (pronase step) and (c) to reduce non-specific binding of probe by acetylation and neutralization of positive charges in the tissues and poly-L-lysine (acetic anhydride step). The pronase digestion conditions have been optimized for floral tissues fixed with either formaldehyde or glutaraldehyde. The signal obtained from pronase treated sections is at least five fold higher than from untreated sections. It may, however, be necessary to alter the length of pronase digestion for use with other tissues or fixatives. The subsequent glycine rinse rapidly inhibits pronase activity.

Protocol 4. Prehybridization treatments

1. Prepare 300 ml of each solution listed below. This is sufficient to treat 2 racks of slides.[a]

Protocol 4. *Continued*

2. Place slides in 25-place stainless steel racks (Raymond A. Lamb) then pass through the following solutions for the indicated times.

Solution	Time
Histoclear[a]	10 min
Histoclear	10 min
100% ethanol[a]	1 min
100% ethanol	30 sec
95% ethanol	30 sec
85% ethanol, 0.85% NaCl	30 sec
70% ethanol, 0.85% NaCl	30 sec
50% ethanol, 0.85% NaCl	30 sec
30% ethanol, 0.85% NaCl	30 sec
0.85% NaCl	2 min
PBS[b]	2 min
Pronase[c] (0.125 mg/ml in 50 mM Tris–HCl, pH 7.5, 5 mM EDTA)	10 min
Glycine (0.2% in PBS)	2 min
PBS	2 min
Formaldehyde[d] (4% in PBS)	10 min
PBS	2 min
PBS (fresh)	2 min
Acetic anhydride[e] (3 ml in 600 ml 0.1 M triethanolamine–HCl, pH 8)	10 min
PBS	2 min
0.85% NaCl	2 min

3. Dehydrate through an ethanol series up to the first 100% ethanol, then wash in fresh 100% ethanol. At this stage, slides can be stored for up to a few hours in a plastic box with a little ethanol at 4°C.

---

[a] Both Histoclear steps and first 100% ethanol must be in glass troughs (Raymond A. Lamb). The rest can be in plastic troughs (Seal-Fresh container 100 mm × 100 mm (Stewart Plastics plc).
[b] PBS, see *Protocol 1*.
[c] Pronase (Sigma type XIV): dissolve at 40 mg/ml in water, and predigest to remove nucleases by incubating at 37°C for 4 h. Store aliquots at −20°C.
[d] Make fresh, as detailed in *Protocol 1*.
[e] Acetic anhydride is unstable in water, thus for acetylation the slide rack must be over a magnetic stirring rod (support by an empty slide rack), stirring rapidly as acetic anhydride is added. Dip the slide rack a couple of times, then stir gently for 10 min. Fresh triethanolamine buffer and acetic anhydride must be used for each rack of slides.

---

4. Hybridization

4.1 Probe type

Single-stranded RNA probes are far superior to double-stranded DNA probes, giving both higher *in situ* signals and lower background noise (3). This is because:

- single-stranded probes cannot reanneal during hybridization;
- the stability of RNA:RNA duplexes is higher than for DNA:RNA duplexes, so sections can be washed at higher stringency;
- non-specifically bound RNA probe can be selectively removed after hybridization by treatment with RNAase A in high salt.

A number of vector systems (e.g. pGEM, pBluescript) are suitable for the synthesis of single-stranded RNA probes from cloned inserts. Transcription reactions are carried out as detailed by the manufacturer of the RNA polymerase used. Once synthesized, probes are subjected to a limited hydrolysis to give fragments of around 150 bases; short probes give higher *in situ* signals (3) as well as a lower background.

The hydrolysis time is calculated from the following equation:

$$t = \frac{Li - Lf}{k \cdot Li \cdot Lf}$$

t = time (min)
k = rate constant ($= 0.11$ kb/min)
Li = initial length (kb)
Lf = final length (kb)

The optimum final length is about 0.15 kb.
Purification and hydrolysis of probes is detailed in *Protocol 5*.

Protocol 5. Probe purification and hydrolysis

1. After the transcription reaction,[a] add 1 μl 100 mg/ml tRNA, 10 units DNAase I (RNAase-free), and H_2O to 100 μl and incubate at 37°C for 10 min.

2. Remove a 1 μl aliquot for measuring incorporation, then extract with 100 μl phenol/chloroform then chloroform.

3. Add an equal vol. 4 M ammonium acetate and 2.5 vol. ethanol, incubate on dry ice for 15 min. Spin, then rinse pellet with 70% ethanol and dry.

4. Resuspend pellet in 50 μl H_2O, add 50 μl carbonate buffer (80 mM $NaHCO_3$, 120 mM Na_2CO_3) and incubate at 60°C for required time.[b]

5. After hydrolysis add 5 μl 10% acetic acid, 0.1 vol. 3 M Na acetate and 2.5 vol. ethanol, chill, spin, and rinse pellet.

6. Resuspend in 20 μl H_2O, count 1 μl.

7. Calculate the % incorporation of nucleotide and the amount of probe synthesized. Add water and formamide to probe to make it 5 times concentrated[c] and 50% in formamide. Store at -20°C.

Protocol 5. *Continued*

^a Transcription reactions, DNAase treatment, and carbonate hydrolysis can be monitored by running a 'cold' reaction in parallel, i.e. using unlabelled rUTP at the same concentration as the other ribonucleotides instead of labelled rUTP. The products are analysed by electrophoresis on a formaldehyde denaturing gel (see Chapter 10).
^b Calculate the hydrolysis time using the equation given in Section 4.1.
^c Usual 1 × probe concentration is 0.1–0.3 ng/μl/kb probe complexity.

4.2 Probe labelling

Riboprobes for *in situ* hybridization can be labelled with a range of isotopes. ³²P-labelled probes have the highest energy, but are not suitable for cellular resolution, and for this reason ³⁵S- or ³H-labelled probes are more often used. I prefer to use ³H-labelled probes since the background is much lower than with ³⁵S-labelled probes and the cellular resolution is higher. rUTP which is double ³H-labelled is available from a number of sources. In a single transcription reaction sufficient probe can be synthesized to hybridize several hundred slides. The probe is stable for several months at −20°C. All probes listed in Section 1 were ³H-labelled and generated relatively high *in situ* signals after 21 day exposures. However, ³⁵S-labelled probes are more sensitive than ³H-labelled probes, and should be considered for use in detecting very low levels of RNA. The modifications to this technique for ³⁵S-labelled probes are detailed in ref. 5.

4.3 Probe concentration

The hybridization signal obtained *in situ* increases as a function of probe concentration up to a saturation point, above which only the background signal increases. In sea urchin embryo sections the saturation point is around 0.3 ng probe/μl hybridization solution/kb probe complexity (7). I find a similar saturation point in plant tissue sections.

4.4 Hybridization conditions

The hybridization conditions detailed in *Protocol 6* have been optimized for homologous RNA probes. Modifications for the use of heterologous probes are discussed in ref. 3.

Protocol 6. Setting up the hybridization

1. Air dry slides and write in pencil on each the probe to be applied.
2. Denature probe (5 × concentrated in 50% formamide) at 80°C for 2 min. Cool on ice then spin briefly.
3. Add 4 vol. hybridization solution (to give final concentrations of 0.3 M NaCl,

Protocol 6. *Continued*

10 mM Tris–HCl, pH 6.8, 10 mM $NaHPO_4$ pH 6.8, 5 mM EDTA, 50% formamide, 10% dextran sulphate, 1 × Denhardts (0.02% each Ficoll, PVP, and BSA), 1 mg/ml tRNA).

4. Apply 40μl hybridization mix to each slide, and spread in a line along the centre of the sections, then carefully apply a 22 × 50 mm coverslip, avoiding bubbles.

5. Place the slides in boxes on tissues soaked in 2 × SSC 50% formamide (1 × SSC is 0.3 M NaCl, 30 mM Na citrate). Seal the boxes with tape and place at 50°C overnight.

5. Post-hybridization treatments and autoradiography

The washing and RNAase treatments detailed in *Protocol 7* are extremely effective at removing non-specifically bound probe, giving mean levels of background noise of less than one silver grain per cell over non-expressing tissues, and a similar low level over the slide (see *Figure 1*). As an alternative to the formamide washes, slides can be washed in 0.1 × SSC at 57°C, but this gives higher background levels. Signals are detected by autoradiography as described in *Protocol 8*.

Protocol 7. Washing procedure

1. Put slides back into stainless steel racks.

2. Immerse in wash buffer (2 × SSC, 50% formamide) degassed before use at 50°C, shake gently for 30 min, after which time coverslips should have fallen off.

3. Change wash buffer and incubate with gentle shaking for 1.5 h. Repeat.

4. Wash for 5 min in NTE (0.5 M NaCl, 10 mM Tris–HCl pH 7.5, 1 mM EDTA) at 37°C. Repeat.

5. Incubate in NTE containing 20 μg/ml RNAase A at 37°C for 30 min.

6. Rinse for 5 min in NTE. Repeat.

7. Wash in wash buffer at 50°C for 1 h.

8. Rinse for 2 min in 1 × SSC at room temperature.

9. Rinse for 30 sec each in 30%, 60%, 90% ethanol (all made 0.3 M in ammonium acetate) then in 95% then 100% ethanol.

10. Air dry.

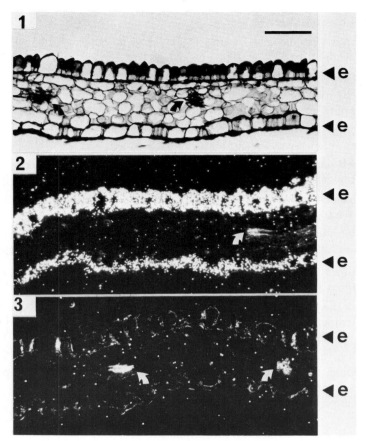

Figure 1. Epidermal-specific expression of chalcone synthase mRNA in petal of *A. majus.* 1. Section of petal of *A. majus,* stained with toluidine blue, e = epidermis, scale bar = 50 μm. 2. Equivalent section hybridized with ³H-labelled antisense chalcone synthase probe and exposed for 21 days. 3. Section hybridized with sense strand chalcone synthase probe. Note: Arrows indicate xylem vessels which are visible under dark field illumination. The epidermal cell walls are also visible.

Protocol 8. Autoradiography

A. *Dipping slides*

Use Kodak NTB-2 or similar emulsion. Use this in a darkroom fitted with a dark red (Wratten No. 2) safelight, which should not shine directly on to the work area.

1. Place the pot of emulsion in a waterbath at 45°C. Leave to melt (approx. 45 min).

Protocol 8. *Continued*

2. Add 5 parts emulsion to 7 parts 1% glycerol in water, also at 45°C. Swirl *gently* to mix.

3. Aliquot into plastic slide mailers (Raymond A. Lamb) (used as dipping chambers). Store aliquots (well wrapped in foil) at 4°C.

4. For dipping, preheat one aliquot for 30 min in a waterbath at 45°C. Invert the slide mailer gently to ensure the emulsion is properly mixed.

5. Dip two blank slides first to remove any bubbles.

6. Dip each slide, withdraw gently, and allow 2–3 sec to drain, then stand in a rack. Return the emulsion to the waterbath every few minutes to keep it warm.

7. When you have dipped all the slides, place the rack of slides in a light-tight box and leave to dry for 1 h.

8. Place the slides in slide boxes with silica gel dessicant, seal with tape, double wrap in aluminium foil, and leave at 4°C for the required exposure time.

B. *Developing slides*

1. Remove the box(es) of slides from the fridge and leave for 1 h to warm up to room temperature. This is extremely important as condensation can reverse 'latent images' in the emulsion.

2. Make up the following solutions, and pre-chill to 14°C which is optimal developing temperature and keeping all solutions at this temperature prevents emulsion cracking. (Solutions cannot be re-used).
   - Developer — Kodak D19
   - Stop — 1% glycerol, 1% acetic acid
   - Fix — 20% sodium thiosulphate

3. In the darkroom, place the slides in a stainless steel rack, then put into developer. Dip rack gently up and down two times, then leave to stand for 2 min. Transfer to stop and agitate the slides gently for 30 sec, then to fix for 5 min. Turn on the lights. Wash the slides in several changes of distilled water for 1 h.

C. *Mounting slides*

If you do not wish to stain the sections, start at step 3.

1. Stain in 0.05% toluidine blue in water for 1 min. (These conditions may have to be varied for different tissue types).

2. Rinse in water.

3. Dehydrate through an ethanol series (in water), see *Protocol 4*.

4. Rinse in Histoclear, then again in fresh Histoclear, drain briefly and add two drops of Depex (BDH) over sections.

Protocol 8. *Continued*

5. Place a coverslip on the slide and press down to squeeze out excess Histoclear/Depex.

6. Leave to dry for at least 1 day.

7. Wash slides in detergent to remove emulsion from the back of the slides.

6. Analysis of results, controls, and artefacts

Autoradiography of sections results in the production of silver grains over the areas where labelled probe has hybridized. If the signal is very strong (e.g. if a ^{32}P-labelled probe has been used or where the gene of interest has a very high level of expression), the silver grains become so dense that they can be seen as black areas under bright field microscopy. In most cases, however, the silver grains are not visible unless viewed under dark field microscopy, where each silver grain shows up as a white speck due to its ability to reflect light (see *Figure 1*).

To test whether the signal observed is due to specific hybridization use a sense-strand probe labelled to the same specific activity; no higher signal relative to background should be observed. A drawback to the use of dark field microscopy is that some plant cell walls also reflect light and thus show up as potentially confusing bright areas. This is especially the case for xylem vessels (see *Figure 1*) and pollen grains. Sections which are stained with toluidine blue also reflect more light than unstained sections and this can cause problems when viewing sections at low magnification. For this reason I prefer not to stain hybridized sections.

A second important control if position-specific gene expression is being investigated, is to show that the areas of the section which do not appear to hybridize do in fact contain RNA which is accessible for hybridization. To demonstrate this one can hybridize adjacent sections using a probe against a constitutively expressed gene. Furthermore, RNA retention can be demonstrated by acridine orange staining (8) although this technique only shows the presence of RNA in cells rich in RNA, such as in meristematic regions and pollen grains and does not establish RNA accessibility.

Use of photographic emulsions leads to a range of possible artefacts and one should consult the literature provided with the emulsion used. One example is chemography where certain chemicals in the section cause the production of silver grains in the emulsion. To test this dip slides which have been through the whole *in situ* procedure except for hybridization with a labelled probe.

Finally, it is important to consider whether differences in silver grain densities reflect real differences in levels of expression. Early work on the sea urchin embryo system indicated that changes in mRNA levels as measured by solution hybridization were accurately reflected by silver grain densities obtained *in situ* (5). Similarly silver grain densities obtained when using several of the pigment biosynthetic genes as probes correlate well with changing pigment distribution over the petal epidermis, suggesting that the technique is at least semi-quantitative within a given tissue type.

Acknowledgements

I wish to thank Cathie Martin and Keith Roberts for supervision and encouragement during this work and critical reading of the manuscript, for which I also thank Angela Hauxwell and Carole Laity. My sincere thanks also go to Sara Wilkinson for typing and to Peter Scott, Barry Allan, and Nigel Hannant for photography. I am supported by a grant from the John Innes Foundation.

References

1. Akam, M. E. (1983). *EMBO J.*, **2**, 2075.
2. Hafen, E., Levine, M., Garber, R. L., and Gehring, W. J. (1983). *EMBO J.*, **2**, 617.
3. Cox, K. H., DeLeon, D. V., Angerer, L. M., and Angerer, R. C. (1984). *Dev. Biol.*, **101**, 485.
4. Ingham, P. W., Howard, K. R., and Ish-Horowicz, D. (1985). *Nature*, **318**, 439.
5. Angerer, L. M. and Angerer, R. C. (1981). *Nucleic Acids Res.*, **9**, 2819.
6. Angerer, L. M., Cox, K. M., and Angerer, R. C. (1987). *Meth. Enzymol.*, **152**, 649.
7. Cox, K. H., Angerer, L. M., Lee, J. J., Davidson, E. E., and Angerer, R. C. (1986). *J. Mol. Biol.*, **188**, 159.
8. McFaden, G. I., Ahluwalia, B., Clarke, A. E., and Fincher, G. B. (1988). *Planta*, **173**, 500.

14

Gene expression using baculoviruses

IAIN R. CAMERON and DAVID H. L. BISHOP

1. Introduction

1.1 The Baculoviridae

The Baculoviridae are a large family of viruses with over 1000 members. The majority have been isolated from insects, predominantly the Lepidoptera but also from Hymenoptera (e.g. sawflies), Diptera (e.g. craneflies), and Coleoptera (e.g. rhinoceros beetle), although examples have also been found in the Crustacea (e.g. shrimps). Baculoviruses have a natural role as regulators of insect populations in the wild, thus their greatest use, until recently, has been as biological control agents for a number of lepidopteran and hymenopteran species which are important pests in agricultural and forestry. Since they are not infectious to vertebrate species or plants they are thought to pose few safety problems. Several reviews are available in which the biology of baculoviruses is comprehensively described (1–3).

The largest Baculovirus subgroup, Nuclear Polyhedrosis Viruses (NPV) includes the type species (named after its host) *Autographa californica* (Ac)NPV which has been extensively studied at the biological and molecular level. Infection with AcNPV, in common with other NPVs, is characterized by the production of a crystalline occlusion body (the polyhedron) whose subunit is an abundantly expressed virus-coded protein termed polyhedrin. Polyhedra contain infectious virus particles and are responsible for horizontal transmission in the wild allowing the virus to persist in the environment, sometimes for many years.

1.2 Baculoviruses as expression vectors

In recent years protein expression systems based on AcNPV (4) have been developed as an alternative to prokaryotic or other eukaryotic expression systems. The main attributes of the system have been well described in a number of recent reviews (5–10). Generation of recombinants is a relatively rapid procedure, high-level expression of biologically-active products is possible, and the system supports a variety of common eukaryotic post-translational modifications including intracellular targetting and secretion (6, 7). AcNPV vectors do not require helper virus, can be readily grown in insect tissue-culture and can be manipulated using standard virological techniques. Analogous protocols have also been developed for another baculovirus, *Bombyx mori* NPV, which exploits silkworm larvae as a vehicle for foreign gene expression (11).

This chapter will concentrate on AcNPV, since this is most applicable to expression in cell culture and is the most widely used system. The polyhedrin gene promoter forms the basis for foreign gene expression in AcNPV. Polyhedrin is expressed in large quantities (often approaching 50% of total cellular protein) late in the virus infection cycle. It is not essential for baculovirus replication and can be dispensed with completely *in vitro*. Replacement of the polyhedrin coding sequence by that of a foreign gene allows expression of that gene. Since the characteristic NPV polyhedra are readily visible in the nuclei of infected cells using a standard plaque counting microscope their presence or absence can provide a visual selection of plaques containing recombinant viruses.

Foreign genes are introduced into AcNPV by way of a transfer vector (*Figure 1*). Several vectors are available, all based on pUC8 carrying the 7.3 kb

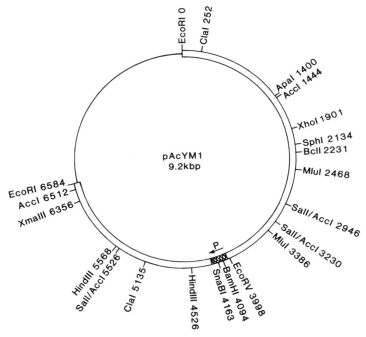

Figure 1. Schematic representation of pAcYM1, a typical baculovirus-based plasmid transfer vector. Restriction sites for enzymes which cut only the baculovirus portion of the vector are shown. This vector consists of the *Eco*RI 'I' fragment of AcNPV (double-lined section) cloned into a modified pUC8 vector (single-lined section). In this vector the coding sequence for polyhedrin has been deleted and replaced by a *Bam*HI linker which acts as the cloning site. The 5' leader and 3' non-coding regions of polyhedrin are shown cross-hatched. The arrow indicates the direction of polyhedrin transcription.

EcoRI 'I' fragment of AcNPV from which the polyhedrin coding sequence has been partially or completely deleted. A unique insertion site is provided immediately downstream of the polyhedrin gene promoter. Such a transfer vector carrying a foreign gene is co-transfected into insect cells with infectious DNA of wild-type AcNPV. Recombinants are selected either by screening for polyhedrin-negative virus plaques or by detection systems, usually based on dot-blot hybridization, using probes for the foreign gene or its product.

Using the AcNPV system, many proteins have been expressed in a biologically active form with correct post-translational modification including glycosylation, phosphorylation, cleavage of polyprotein precursors, signal peptide processing and secretion, and fatty-acylation. In addition intracellular targeting can also be achieved. While the level of expression cannot be guaranteed to match that of polyhedrin the large number of proteins expressed so far may allow some prediction of likely yield. For many genes high level expression (10–50% of cellular protein) is obtained. Foreign proteins expressed using AcNPV vectors include viral antigens for vaccine production, DNA and receptor binding proteins and enzymes while expressed plant gene products include french bean phaseolin and patatin.

In this chapter the methods for generation of a recombinant baculovirus expressing a foreign gene product are described. This process involves a number of standard cloning and detection methods which can be found in a variety of practical molecular biology guides (e.g. ref. 19).

2. Cells, viruses, and transfer vectors

2.1 *Spodoptera frugiperda* cells

Spodoptera frugiperda IPLB-Sf21-AE cells (Sf21) (12) are an insect ovarian cell line derived from pupae of the fall armyworm. A clonal isolate, Sf9 (8), is available from the American Type Culture Collection (No CRL 1711). Both cell lines have been used successfully for expression from baculovirus vectors and parameters for growth of these cells are given in *Table 1*.

2.2 *Autographa californica* Nuclear Polyhedrosis Virus

Autographa californica Nuclear Polyhedrosis Virus (AcNPV) is the prototype virus of the family Baculoviridae. It is a double-stranded DNA virus with a circular genome of 130 kb. Commonly used strains are denoted E2 and C6.

2.3 Some commonly used transfer vectors

The commonly used transfer vectors have been well described in a review by Luckow and Summers (6). Briefly they can be considered in four groups; in all

Table 1. *Spodoptera frugiperda* cells—parameters in tissue culture

Cell name	IPLB-Sf21-AE or subclone Sf9 (ATCC# *CRL 1711*)
Medium	TC100 supplemented with 5% or 10% FCS
Doubling time	18–24 hours
Incubation temperature	28°C for rapid growth and virus work 21°C for routine passage
Type of culture	Monolayer or suspension
Seeding density	2×10^4 cells/cm^2 (monolayer culture) 10^5 cells/ml (suspension culture)
Confluent monolayer density	2×10^5 cells/cm^2 (e.g. 1.5×10^6 cells/35 mm plate)
Suspension culture maximum density	$2–4 \times 10^6$ cells/ml
Passages used for virus culture	145–180

cases expression is driven by the intact promoter from the polyhedrin gene upstream of a full length polyhedrin leader sequence:

- pAc373 and pAcYM1 have a *Bam*HI cloning site immediately downstream of the leader sequence and the foreign gene must supply its own AUG codon.

- The cloning site in pAc360 and similar vectors, is preceeded by several polyhedrin-derived codons (including the AUG). These vectors produce a fusion product which may be useful in instances where the foreign protein is unstable.

- Secretion vectors (e.g. pAcJM105) containing a signal peptide allow secretion of the product and allow study of protein targeting within the cell.

- The newest class of vectors are multiple expression vectors (13) in which duplicated baculovirus promoters (not necessarily polyhedrin) allow expression of several foreign genes, one of which may be polyhedrin (e.g. pAcVC2). A feasible alternative is coinfection with several recombinant viruses which has the advantage that the relative contributions of the two recombinants can be varied.

3. Culture of *Spodoptera frugiperda* (Sf) cells

Sf cells are routinely cultured in TC100 or Graces medium (FLOW or GIBCO) supplemented with 5–10% fetal calf serum (FCS). Antibiotics, penicillin (50 units/ml), streptomycin (50 µg/ml), and kanamycin (50 units/ml) and the antimycotic, fungizone (2.5 µg/ml) can be added if desired and may be particularly helpful during plaque assays. It is essential that cells are maintained in a healthy state (> 98% viable) for all manipulations with baculoviruses. This is routinely assessed by counting cells in the presence of trypan blue (0.05% w/v) each time they are passaged.

3.1 Monolayer culture

Cultures of Sf cells are maintained by passage every 5 or 6 days at 21°C in glass or plastic flasks. Virus infected cells are kept at 28°C. Confluent cells are gently scraped or shaken into the medium (trypsin is not required) and subcultured at a 1:10 dilution (2×10^6 cells/cm^2). Sf cells have no minimum plating density therefore lower densities of cells can be used to maintain stocks for longer periods (up to 10 days) at room temperature. For experimental work with viruses, 35 mm Petri dishes are seeded with 1.5×10^6 cells in 2 ml of medium, the plates can be used after 1–2 h or kept overnight at 28°C. Cells can be infected when their density reaches 5×10^6 cells/ml.

3.2 Suspension culture

Sf cells readily adapt over 1–2 passages to suspension culture and this is used for large scale production of virus stocks or expression of proteins by recombinant viruses. Round bottom flasks are seeded with 10^5 cells/ml of medium and stirred at 100 r.p.m. using a magnetic stirring bar. To maintain a good oxygen supply, flasks should be no more than half full and cultures should not normally exceed 500 ml. Several flasks will suffice for most experimental work.

3.3 Cell storage

Exponentially growing cells are diluted to $1–4 \times 10^6$ cells/ml in culture medium containing 10% DMSO. Aliquots (0.5–1 ml) are frozen and stored in liquid nitrogen. Cells are revived by rapid thawing and diluted to 5 ml with medium. After overnight incubation at 28°C the medium is renewed and the cells can be cultured normally.

4. *In vitro* manipulation of AcNPV baculovirus

4.1 Titration of AcNPV baculovirus in cell culture

The production of clearly defined virus plaques is central to many of the procedures described here. It is essential to know accurately the titres of virus stocks so that the appropriate multiplicity of infection (usually 10 pfu/cell) can be used to achieve optimum expression. When screening for recombinant viruses polyhedrin-negative plaques must be distinguished from those containing polyhedra and then purified by repeated plaque-picking and titration.

Protocol 1. Titration of AcNPV baculovirus

1. Harvest and count actively growing Sf cells (>98% viable—Section 3.3).
2. Dilute the cells to 7.5×10^6 cells/ml in TC100 and dispense 2 ml into 35 mm tissue culture Petri dishes. Incubate 1–2 h at 28°C.

Protocol 1. *Continued*

3. Prepare serial 10-fold dilutions (0.5–1 ml) of virus in PBS or TC100 (12) to cover the range of the expected titre.

4. Remove the medium from the cell monolayers and gently inoculate 100 μl of virus dilution into the centre of the dish. Rock gently to distribute the virus.

5. Incubate at room temperature for 1 h.

6. Remove the inoculum from the edge of a tilted dish with a Pasteur pipette.

7. To prepare agarose overlay while adsorbing virus to the cells mix 3% LGT-agarose (previously dissolved and sterilized by autoclaving or in a microwave and cooled to 37°C) with an equal volume of TC100 + 5% FCS at 37°C. Keep the overlay at 37°C until use.

8. Add 1.5 ml of agarose overlay to the side of the dish and allow it to spread slowly over the cells. Once set (approx. 20 min) add 1 ml of TC100 + 5% FCS containing antibiotics as a liquid feeder layer.

9. Incubate the plates in a humid environment (e.g. clear polystyrene sandwich boxes) at 28°C for 3–4 days or until plaques are well formed.

10. Observe plaques directly by viewing a dish, side-lit with an intense light source, against a dark background. Alternatively stain the cells by adding 1 ml of neutral red dye (0.01% w/v in PBS) to the overlay. Incubate for 2–4 h at 28°C then drain the liquid by inverting the dish over a beaker, blot residual stain on a clean paper towel and leave the dishes for 5–18 h at room temperature. Plaques will appear clear against a red background of uninfected cells and can be counted by eye over a light box.

11. To isolate recombinant virus perform three rounds of plaque purification from well separated plaques (see *Protocol 2*). Select plaques using a suitable inverted microscope to observe the plaques in detail (e.g. for the presence or absence of polyhedra).

4.2 Propagation of AcNPV in *Spodoptera frugiperda* cells

Protocol 2 describes the amplification of virus from single plaque into a large suspension culture and illustrates the techniques required for virus purification.

Protocol 2. Propagation of AcNPV

1. Remove a plug of agarose from above a clear, well isolated plaque using a sterile Pasteur pipette. Disperse the agarose in 0.5 ml of medium containing antibiotics (Section 3). The expected titre is 10^4–10^5 pfu/ml.

2. Inoculate 100 μl of each plaque isolate on to 1.5×10^6 cells in a 35 mm Petri dish from which the medium has been removed, and allow the virus to adsorb for 1 h.

Protocol 2. *Continued*

3. Add 2 ml of TC100 + 5% FCS and incubate 3–4 days at 28°C until cytopathic effects are seen. The medium from this plate will contain 10^6–10^7 pfu/ml and can be used to genetate high titre virus stocks.

4. Seed an 80 cm^2 tissue culture flask with 5×10^6 cells, inoculate with virus at an m.o.i. of 0.1–0.5 pfu/cell and incubate with 10 ml of medium for 3–4 days. The virus titre should now be 5×10^7–10^8 pfu/ml.

5. Produce a very high titre stock of virus by inoculating a 500 μl suspension culture containing 5×10^5 cells/ml with 0.1–0.5 pfu/cell. Virus can be purified after 3–4 days (see *Protocol 3*).

4.3 Purification of infectious AcNPV DNA

Successful transfection of insect cells requires DNA of high quality and purity and care is required to avoid loss of infectivity through shearing. DNA must be prepared from virus particles of an appropriate genotype (normally wild-type). Some procedures may require DNA of a polyhedrin-negative or other special genotype. Purified virus particles can be stored at 4°C and DNA purified from them in small batches (*Protocol 3*). Infectious DNA is normally used within 2–4 weeks of preparation.

Protocol 3. Purification of AcNPV/DNA

1. Inoculate a 500 ml suspensin culture (5×10^5 cells/ml) with 0.1–0.5 pfu/ml of AcNPV and incubate at 28°C for 3 days.

2. Remove the cells by low speed centrifugation (1000 g 10 min).

3. Transfer the supernatant to 50 ml tubes (e.g. oak-ridge) and pellet the virus particles at 75 000 g for 1 h at 4°C.

4. Resuspend the pellet gently after soaking overnight in 1–2 ml of TE^a buffer, and remove any remaining cellular debris at 1000 g for 10 min.

5. Layer the virus on to a 10% to 50% (w/v) discontinuous sucrose gradient and centrifuge at 100 000 g for 1 h at 4°C in a swing out rotor (e.g. Beckman SW-41).

6. Remove the thick, white band from above the 50% sucrose layer with a pipette or by downward displacement. Dilute the virus 4-fold with TE and pellet at 75 000 g for 1 h at 4°C.

7. Resuspend the pellet in approximately 2 ml of TE and store at 4°C.

8. Mix 200 μl of virus particles with 200 μl of TE and add 100 μl of 20% sarkosyl in TE, incubate at 60°C for 60 min to lyse the virus.

9. Layer the lysate on to 5 ml of 54% (w/w) CsCl in TE containing 50 μg/ml ethidium bromide and centrifuge at 210 000 g for 18 h in, for example, an SW41 rotor.

Protocol 3. *Continued*

10. Harvest the DNA from the tube by side-puncture or by downward displacement. Collect both the lower (supercoiled) and upper (relaxed) DNA bands. These can be pooled since both are infectious.

11. Remove the ethidium bromide by gentle extraction with butan-1-ol (repeat 4 times) followed by extensive dialysis against 5–6 changes of approx. 1000 vol. TE at 4°C.

12. Measure the DNA concentration by absorbance at 260 nm and store at 4°C. *Do not freeze since this destroys infectivity.* The yield shoud be around 10–20 μg.

a TE buffer is 10 mM Tris, 1 mM EDTA, pH 7.5.

---

5. Generation of recombinant AcNPV baculoviruses

The procedures for generation of a recombinant baculovirus expressing a foreign gene are summarized in *Figure 2*, and discussed in Section 5.1.

5.1 Insertion of foreign genes into a transfer vector

The available transfer vectors have a unique cloning region into which the foreign gene can be inserted by standard techniques. In the case of pAcYM1 or pAc373 the single *Bam*H1 site is used. The foreign gene can be cloned readily if it has compatible sticky ends, otherwise both the transfer vector and foreign gene may be rendered blunt-ended using mung-bean nuclease. If the foreign gene has a significant amount of 5′ non-coding DNA then as much as possible should be removed (e.g. with Bal31), particularly if it is GC-rich; in contrast, the amount of 3′ non-coding sequence appears to have little effect on expression. Normally the foreign gene should have its own initiation and termination codons although certain vectors retain the polyhedrin ATG (e.g. pAcRP14), or contain a leader sequence (e.g. from polyhedrin in pAc360 or VSV glycoprotein in pAcJM25) which results in the expression of a fusion polypeptide.

After insertion of the foreign gene into the transfer vector it is prudent to sequence the junctions to confirm the expected sequence arrangement. Either Maxam and Gilbert chemical sequencing using a suitable restriction site in the insert and the EcoRV site 90 bp upstream of the polyhedrin ATG or Sanger dideoxy sequencing using appropriate primers and double-stranded plasmid DNA may be used.

Purify the final construct by CsCl density gradient centrifugation and keep the DNA solution as sterile as possible. Store at 4°C ready for transfection into insect cells.

5.2 Transfection of vector and virus DNA into insect cells

Currently the calcium phosphate co-precipitation technique is used to introduce virus and vector DNA into Sf cells according to *Protocol 4*.

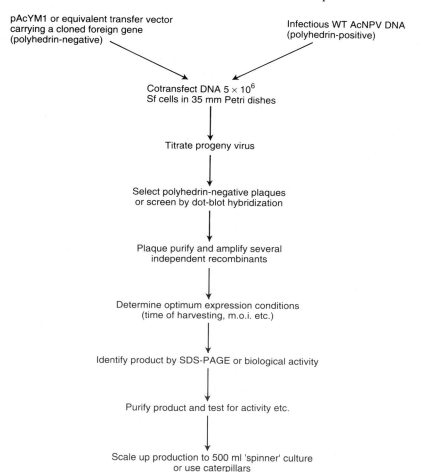

pAcYM1 or equivalent transfer vector
carrying a cloned foreign gene
(polyhedrin-negative)

Infectious WT AcNPV DNA
(polyhedrin-positive)

Cotransfect DNA 5×10^6
Sf cells in 35 mm Petri dishes

Titrate progeny virus

Select polyhedrin-negative plaques
or screen by dot-blot hybridization

Plaque purify and amplify several
independent recombinants

Determine optimum expression conditions
(time of harvesting, m.o.i. etc.)

Identify product by SDS-PAGE or biological activity

Purify product and test for activity etc.

Scale up production to 500 ml 'spinner' culture
or use caterpillars

Figure 2. Diagram of the steps involved in the construction of a recombinant baculovirus and subsequent expression of a foreign protein using *Spodoptera frugiperda* cells.

Protocol 4. Transfection of DNA into insect cells

1. Seed sufficient 35 mm Petri dishes with 1.5×10^6 Sf cells.

2. Mix:

- $2 \times \text{HeBS}^a$ 500 μl
- 100 mM Glucose   100 μl
- AcNPV DNA   1 μg
- Transfer vector plasmid DNA   5 μg
- H_2O to 935 μl

Protocol 4. *Continued*

3. Rapidly add 65 μl of 2 M CaCl$_2$ and immediately vortex vigorously. Stand at room temperature for 30 min during which time a fine precipitate should form.

4. Remove the medium from the cell monolayers and gently pour on the DNA precipitate. Incubate at room temperature for 1 h.

5. Add 2 ml of TC100 + 5% FCS and incubate at 28°C for 4 h.

6. Remove the medium + DNA and add 2 ml of fresh medium. Continue incubating for 2–3 days.

7. Observe the cells using an inverted miroscope. The presence of polyhedra confirms that transformation has been successful. Inclusion of a control transfection using AcNPV DNA only may be useful for this purpose.

8. Harvest the cells and medium and pellet the cells at 1000 g for 10 min.

9. Titrate the supernatant virus (*Protocol 1*) using dilutions ranging from 10^{-1} to 10^{-4}. Store undiluted virus at 4°C; this is stable for at least several months.

a 2 × HeBS (Hepes Buffered Saline) = 40 mM HEPES, pH 7.05, 2 mM Na$_2$HPO$_4$, 10 mM KCl, 250 mM NaCl. Filter sterilise and store in small aliquots.

5.3 Selection of recombinant viruses

The identification of recombinant viruses lacking the polyhedrin gene can be achieved by direct observation of plaques using an inverted microscope (*Protocol 5*). With practice this method can be very successful. Alternatively screening methods using appropriate DNA or antibody probes may be used on virus infected cells grown in microtitre plates (*Protocol 5*).

Protocol 5. Selection of recombinant virus

A. *Visual identification of virus plaques*

1. Infect 10 to 15 35 mm dishes of Sf cells with the supernatant from a transfection experiment using 10^{-1} to 10^{-3} dilutions or the optimum dilution if known from past experience.

2. Incubate for 3 days at 28°C and stain with neutral red dye (*Protocol 1*).

3. Identify polyhedrin-negative plaques using a low-power lens on an inverted microscope. These plaques appear clear and lack the characteristic refractile polyhedral inclusions in the nucleus. With practice it should be possible to identify polyhedrin-negative plaques with relative ease.

4. Pick polyhedrin-negative plaques into 0.5 ml of TC100 and retitrate (usually

Protocol 5. *Continued*

3 times) at 10^0 to 10^{-2} dilutions until only polyhedrin-negative virus is obtained in the assay.

B. *Identification by dot-blot hybridization*

1. Titrate the transfection supernatant (*Protocol 1*) and dilute to 50 pfu/ml.
2. Seed a 96-well tissue culture plate with 10^4 cells per well and incubate at 28°C for 1 h.
3. Remove the medium and add 200 μl (10 pfu) of diluted transfection mix. Incubate in a humid box at 28°C for 72 h.
4. Transfer the culture fluid to a fresh 96-well plate and store sealed at 4°C.
5. Further processing of the cells remaining in the wells depends on the detection method chosen.
   - For DNA dot-blotting resuspend the cells in lysis buffer according to Kafatos *et al.* (13).
   - For Western dot-blotting, freeze and thaw cells in a buffer such as RIPA (1% Triton X-100, 1% sodium deoxycholate, 0.5 M NaCl, 0.05 M NaCl, 0.05 M Tris–HCl, pH 7.4, 0.01 M EDTA, and 0.1% SDS).
   - Perform ELISA, or an appropriate activity assay, on the cells or virus-containing supernatant.

---

6. Expression of foreign genes from baculoviruses

Once a recombinant virus has been identified, expression of the foreign protein can often be demonstrated by SDS-PAGE of infected cells. Several independent isolates (usually five) should be tested to allow for variation in expression levels. Cells should be infected with 10 pfu/cell and harvested at various times post-infection to investigate the level of foreign protein expression by staining the proteins resolved by SDS-PAGE. Radioactive labelling may be necessary in order to detect the product and may also be used to monitor product processing by the use of, for example, ^{14}C-mannose for glycosylation or ^{32}P NTP for phosphorylation.

When expression has been established large scale production of the foreign protein can be attempted using one or more 500 ml suspension cultures. Cells are grown to a density of 5×10^5 per ml, concentrated by gentle centrifugation to 10^7 cells/ml, infected with 10 pfu recombinant virus/cell, shaken gently for 1 h and rediluted to 10^6 cells/ml using fresh TC100 + 10% FCS. Protein can be purified from the cells or supernatant after stirring at 28°C for the appropriate time, usually 1 to 3 days but determined experimentally for each recombinant.

Yields of foreign protein from 3×10^9 cells using this system have ranged from 0.2 mg/litre to 500 mg/litre (equal to that of polyhedrin and equivalent to 50% of cellular protein). Commonly 5–10 mg is recovered with little difficulty, but much depends on the characteristics of the foreign protein. It may be necessary to concentrate secreted proteins from the culture medium, this will require the use of serum free medium (e.g. EX-CELL 400 from JR Scientific [Seralab]). Contamination with serum proteins can also be avoided by incubating the cells in TC100 without FCS for 24 h prior to harvesting, however this method will not be suitable for products expressed over an extended period of culture. An alternative means of expression is to infect larvae of an appropriate insect host (*Trichoplusia ni* in the case of AcNPV) with the recombinant virus and purify the foreign protein from homogenized larvae. Although this method has been used successfully (15) methodologies for this procedure have not been standardized.

Where yield is low and cannot be maximized by varying m.o.i. ot time of harvesting, further considerations which may improve expression include:

(a) If no expression can be detected.

- Check the construction of the transfer vector and recombinant virus by restriction mapping and DNA sequencing at the foreign gene insertion site.

- Use Southern blotting techniques to determine that the foreign gene has been inserted into the recombinant virus and is correctly located.

- Use northern blotting techniques to check that the gene is being transcribed, and to determine its time course.

- Check the consensus sequence around the ATG codon. Animal genes normally follow the Kozak rules (16). For plant genes it may be necessary to adjust the sequence around the ATG.

(b) If the construct appears to be correct or if expression can be detected but is poor consider modifying the transfer vector.

- The amount of 5' non-coding DNA in the insert can be reduced although this does not always have any effect on expression.

- Out-of-frame ATG codons upstream of the desired translational start site should be deleted as these could interfere with expression.

- Consider using a fusion vector such as pAc360 (Section 2.3) or adding a signal sequence to aid protein secretion. The signal sequence from VSV may be suitable for this purpose. For membrane proteins, changing a signal sequence (or deleting it for internal accumulation of the product) may increase expression levels.

- Check for spurious polyhedrin RNA initiation sites in the foreign gene and its complement.

Once satisfactory expression has been obtained with the system it should be possible to produce at least 5–10 mg of virtually any protein. In our experience it is almost always possible to achieve some expression.

6.1 Expression of hepatitis B proteins in a baculovirus vector

The expression of hepatitis B virus surface (HBsAg), core (HBcAg), and pre-core

(HBpcAg) antigens in a baculovirus vector provides a good example of the use of the AcNPV system (15). DNA containing the coding sequences of HBsAg, HBcAg, or HBpcAg were inserted individually into the transfer vector pAcYM1 and recombinant virus were produced as described above (Section 5). In addition dual expression vectors were constructed to express either, HBsAg and HBcAg, or, HBsAg and polyhedrin. A Coomassie-blue stained gel of infected-cell extracts demonstrates expression of HBsAg, HBcAg, HBpcAg, and polyhedrin at 2 days

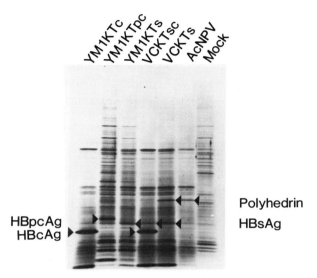

Figure 3. Coomassie blue stained gel showing expression of hepatitis B antigens from recombinant baculoviruses. Recombinants YM1KTc, YM1KTpc, and YM1KTs express HBcAg, HBpcAg, and HBsAg antigens respectively. VCKTsc and VCKTs are dual expression vectors expressing HBsAg + HBcAg and HBsAg + polyhedrin respectively. AcNPV is Sf cells infected in parallel with WT virus. Mock infected cells are also shown. Cell extracts were obtained at 2 days post-infection. The positions of HBcAg, HBpcAg, HBsAg, and polyhedrin proteins are indicated. (Data reproduced from ref. 15).

post-infection (*Figure 3*). Expression of the predominantly cell-associated HBcAg is estimated at 40% of cell protein, HBsAg and HBpcAg are estimated to be 2% and 5–10% respectively. The apparent low level of polyhedrin is due to the low m.o.i. (1 pfu/cell) used; at 10 pfu/cell polyhedrin will reach 40% of cellular protein at 2 days post-infection. The time course of expression of HBsAg (*Figure 4*) shows the total yield to be approximately 0.4 mg per litre of 10^9 cells, and demonstrates that whereas total synthesis peaks at 2–3 days post-infection the relative contributions of secreted and cell-associated HBsAg continue to alter after this. When the HBsAg polyhedrin dual expression vector was used to infect *T.ni* larvae between 2 and 4 μg of HBsAg were purified from each dead larvae. In comparison polyhedrin-negative virus has a low infectivity for insects.

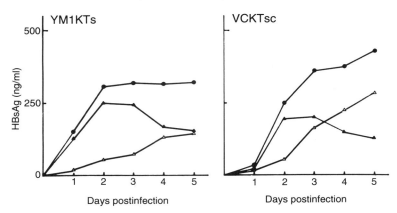

Figure 4. Time course of expression of HBsAg from vectors YM1KTs and VCKTsc. Samples (5 ml) were taken from a 500 ml suspension culture daily and cells were separated by centrifugation. HBsAg content in the cells and supernatant were assayed by radioimmunoassay. (Data reproduced from ref. 15).
●—● = total yield
▲—▲ = cell associated protein
△—△ = secreted protein

6.2 Systems for multiple expression of foreign genes

Expression of several proteins in the same cell can be achieved by the simple expedient of co-infecting the cell with multiple recombinant viruses (17). However, a more sophisticated approach has been developed involving a transfer vector carrying two copies of the polyhedrin gene promoter (13), this however may be more limiting in practice and can involve as much work as the creation of two separate recombinants. Multiple expression allows the interaction of the expressed proteins to be studied *in vitro* (18). If one of the expressed proteins is polyhedrin then this may facilitate infection of insects as a means of protein production (Section 6.1; ref. 15). Multiple expression vectors are still under development and new vectors incorporating promoters from other baculovirus genes are to be expected in the future. Some of these may allow expression earlier in the virus infection cycle making possible temporally regulated expression of foreign genes. Baculoviral expression systems hold great promise for expression of plant proteins.

Acknowledgements

The authors would like to thank Kazu Takehara for the use of his data on hepatitis B antigen expression in baculoviruses and Chris Hatton for photographic work.

References

1. Kelly, D. C. (1982). *J. Gen. Virol.*, **63**, 1.
2. Doerfler, W. and Bohm, P. (ed.) (1986). *Cur. Top. Micr. Imm.*, **131**, Springer, Berlin.
3. Granados, R. R. and Federici, B. A. (ed.) (1986). *The biology of baculoviruses*, Vol. 1. *Biological properties and molecular biology*. CRC Press, Boca Raton, Florida.
4. Smith, G. E., Fraser, M. J., and Summers, M. D. (1983). *J. Virol.*, **46**, 584.
5. Miller, D. W., Safer, P., and Miller, L. K. (1986). In *Genetic engineering: principles and methods*, Vol. 8 (ed. J. K. Setlow and A. Hollaender), p. 277. Plenum, New York.
6. Luckow, V. A. and Summers, M. D. (1988). *Bio/Technology*, **6**, 47.
7. Miller, L. K. (1988). *Annu. Rev. Microbiol.*, **42**, 177.
8. Summers, M. D. and Smith, G. E. (1987). *A manual of methods for baculovirus vectors and insect cell culture procedures: Bulletin No. 1555*. Texas Agricultural Experimental Station.
9. Cameron, I. R., Possee, R. D., and Bishop, D. H. L. (1989). *Trends in Biotechnology*, **7**, 66.
10. Bailey, M. J. and Possee, R. D. (1991). In *Methods in molecular biology*, vol. 7 (ed. E. J. Murray), p. 147. Humana Press, New York.
11. Maeda, S., Kawai, T., Obinata, M., Fujiwara, H., Saeki, Y., Sato, Y., and Furusawa, M. (1985). *Nature*, **315**, 592.
12. Vaughn, J. L., Goodwin, R. H., Tompkins, G. L., and McCawley, P. (1977). *In Vitro*, **13**, 213.
13. Emery, V. C. and Bishop, D. H. L. (1987). *Protein Eng.*, **1**, 359.
14. Kafatos, F. C., Jones, C. W., and Efstratiadis, A. (1979). *Nucleic Acids Res.*, **7**, 1541.
15. Takehara, K., Ireland, D., and Bishop, D. H. L. (1988). *J. Gen. Virol.*, **69**, 2763.
16. Kozak, M. (1983). *Microbiol. Rev.*, **45**, 1
17. St Angelo, C., Smith, G. E., Summers, M. D., and Krug, R. M. (1987). *J. Virol.*, **61**, 361.
18. Urakawa, T., Ferguson, M., Minor, P. D., Cooper, J., Sullivan, M., Almond, J. W., and Bishop, D. H. L. (1989). *J. Gen. Virol.*, **70**, 1433.
19. Sambrook, J., Fritsch, E. F., and Maniatis, T. (ed.) (1989). In *Molecular cloning: a laboratory manual* (2nd edn). Cold Spring Harbour Press, New York.

Characterized defence response genes

RICHARD A. DIXON

Table 1 lists a range of plant defence response cDNA clones as reported in the literature as of September 1990. Reports of successful heterologous cross-hybridizations are included in column 5 for workers who are interested in exploring the use of heterologous probes for those species from which the sequence has not yet been cloned. Some reports are personal communications to the author and suitability for use with a plant species other than that of origin should therefore be taken as a suggestion only. Genes from sources marked with a * are, as far as is known, not involved in induced defence in the species from which they were isolated, but have been identified as defence-response components in other species.

A similar list of reported genomic clones is given in *Table 2*.

Table 1. cDNAs for plant defence response genes

Gene encoding	Source	Vector	Designation	Comments	References
Enzymes of phytoalexin biosynthesis					
Phenylalanine ammonia lyase	*Ipomoea batatas*	pUC8	pPALO2	2.12 kb cDNA. mRNA from wounded root tissue.	70
	Oryza sativa	λgt11	λCP-1	2.5 kb cDNA. mRNA from leaves	46
	Phaseolus vulgaris	pAT153	pPAL5	1.7 kb cDNA. mRNA from elicitor-induced cell suspension cultures. Cross-hybridization with *Petroselinum* PAL and other legume PALs.	16
	Petroselinum crispum	pBR322		From elicitor-induced cell suspension cultures. Cross-hybridization with bean and potato PALs.	38, 64
	Solanum tuberosum	λgt10		1.5 kb cDNA. mRNA from elicitor-induced cell suspension cultures. Cross-hybridization with bean and *Petroselinum* PALs. Contains internal Eco R1 site; 0.6 and 0.9 kb subclones in pUC9.	21
4-Coumarate: CoA ligase	*Petroselinum crispum*	pBR322	pc4CL	mRNA from elicitor-induced or UV-induced cell suspension cultures. Does not cross-hybridize with potato 4CL.	32
	Solanum tuberosum	λgt11		2.03 kb cDNA. mRNA from elicitor-induced cell suspension cultures.	21
Chalcone synthase	*Matthiola incana**	λNM1149	pcM12	1.38 kb cDNA. mRNA from pigmented petals.	17
	Medicago sativa	λZAPII		cDNAs encoding individual members of the alfalfa CHS multigene family.	12
	Petroselinum crispum	pBR322	pLF15	1.5 kb cDNA	54
	*Petunia hybrida**	pUC9	VIP50	1.5 kb full length cDNA, mRNA from petals	31

192

Enzyme	Species	Vector	Clone	Description	Ref.
	Phaseolus vulgaris	pBR325	pCHS1	1.4 kb cDNA. mRNA from elicitor-induced cell suspension cultures. Cross-hybridization to CHS from a number of species.[a]	56
	Phaseolus vulgaris	pBR325, with various subclones in pSP64 or pSP65		Subclones specific for individual bean CHS genes.	57
	Pisum sativum	λNM1149	pCHS1/2	cDNAs for 2 members of the pea CHS multigene family.	22
Chalcone isomerase	Petunia hybrida*	λgt11	CHI-A	One 0.75 kb cDNA. mRNA from corolla tissue.	73
	Phaseolus vulgaris	λgt11, subcloned in pSP65	pCHI-1	0.86 kb cDNA. mRNA from elicitor-induced cell suspension cultures.	1, 43
Stilbene synthase	Arachis hypogaea	λNM1149	pGSC1, pGSC2	Two cDNA clones. mRNA from cell suspension cultures.	60
3-Hydroxymethyl-3-glutaryl coenzyme A reductase	Arabidopsis thaliana*	λgt10, subcloned in pUC19	HMG1		36
	Arabidopsis thaliana*	λgt10	λCAT1/2	Two overlapping cDNA clones. mRNA from leaves.	4
	Lycopersicon esculentum	λgt10		mRNA from fruit; possibly more than one gene.	48
	Solanum tuberosum	PCR[2], subcloned in pSP72	—	From elicitor-treated tuber tissue amplified by PCR.[b] Two distinct cDNAs (88% homologous).	67
Casbene synthase	Ricinus communis	λgt11	pCS4	Partial cDNA sequence. mRNA induced by pectic fragment elicitors.	37

Enzymes of cell wall phenolic biosynthesis

Enzyme	Species	Vector	Clone	Description	Ref.
Cinnamyl alcohol dehydrogenase	Phaseolus vulgaris	λgt11, subcloned in pUC19	pCAD4a	Single cDNA. mRNA from elicitor-induced cell suspension cultures. Internal EcoR1 site. Second gene involved in xylogenesis?	74

Table 1. *Continued*

Gene encoding	Source	Vector	Designation	Comments	References
Lignin peroxidase	*Nicotiana tabacum*	λgt11		1.25 kb cDNA. mRNA from leaf tissues; 4 genes.	33
Hydroxyproline rich glycoproteins					
Hydroxyproline-rich glycoprotein	*Nicotiana tabacum**	pUN121	—	cDNA for message induced by cytokinin.	44
	Phaseolus vulgaris	pUC19	Hyp 3.6, Hyp 2.13, Hyp 4.1	Three different cDNAs (0.7–1.6 kb) corresponding to different length transcripts. 3 genes, differential induction.	5
	Phaseolus vulgaris	pUC19	Hyp 2.11	Wound-induced cDNA, single gene (different from above).	59
Hydrolases and 'PR' proteins					
Chitinase	*Cucumis sativus*	λZAP		cDNA for extracellular, acidic chitinase. No homology to bean or tobacco chitinases. One gene.	45
	Hordeum vulgare	Not stated		0.56 kb cDNA. mRNA from aleurone layers.	68
	Nicotiana tabacum	pBR322		cDNA for hormonally-regulated message.	61
	Nicotiana tabacum	pUN121		cDNA for message induced by cytokinin.	44
	Nicotiana tabacum	pUC9		cDNAs for virus-induced acidic and basic chitinases; acidic and basic each encoded by at least 2 genes.	26
	Nicotiana tabacum	λOngC		cDNAs for two isoforms of acidic endochitinase PR proteins P and Q.	51
	Phaseolus vulgaris	λgt11		cDNAs for message induced by ethylene in leaves.	3
	Phaseolus vulgaris	λgt11, subclone pCHT12.1 in pBluescript		0.65 kb cDNA fragment. Message from elicitor-induced cell suspension cultures.	23

194

	Species	Vector	Clone	Description	Ref.
1,3-β-D-glucanase	Solanum tuberosum	Not stated	CCH4	1.09 kb cDNA with homology to bean and tobacco chitinases.	34
	Glycine max	λgt11	pEG488	1.26 kb cDNA. mRNA from cotyledons. Induced by ethylene.	69
	Hordeum vulgare*	pUC9		cDNA for germination-induced message (1,3:1,4-activity).	19
	Medicago sativa	λZAPII		Full length cDNA encoding an acidic glucanase. Small gene family.	39
	Nicotiana plumbaginifolia	pUC18	pBEG	1.29 kb cDNA for homonally-regulated message. 75% homologous to N. tabacum; 52% homologous to Hordeum vulgare (1,3:1,4 glucanase).	13
	Nicotiana tabacum	pBR322		cDNA for mRNA whose production is inhibited by both auxin and cytokinin.	47
	Nicotiana tabacum	pBR322		cDNA sequences indicate at least 4 transcriptionally active genes.	63
	Phaseolus vulgaris	pUC19		cDNA for message from elicitor-induced cell suspension cultures. Cross-hybridizes to alfalfa genomic sequences. Single gene.	15
'Pathogenesis-related' protein	Nicotiana tabacum	pUC9		cDNAs encoding PR-1a, PR-1b, and PR-1c gene transcripts. Virus-induced.	7, 25
	Nicotiana tabacum	pUC12		cDNAs encoding PR-1a, PR-1b, and PR-1c gene transcripts. Virus-induced.	52
	Nicotiana tabacum	pUC8		PR-1a and PR-1b cDNAs. Multigene families, at least 2 PR-1as and 4 PR-1bs.	40
	Nicotiana tabacum	pUN121		PR-1 and 'PR-1-like' cDNAs from message induced by cytokinin.	44
	Nicotiana tabacum	λgt11		PR-1a, PR-1b, and PR1c cDNAs. Induced by virus infection.	11

Table 1. *Continued*

Gene encoding	Source	Vector	Designation	Comments	References
	Nicotiana tabacum		various	Basic chitinase and glucanase and osmotin induced by viral infection. cDNAs isolated from thin cell layer explants initiating floral development.	49
	Nicotiana tabacum	pUC9		cDNA for thaumatin-like protein.	27
	Phaseolus vulgaris	λgt11	PV PR1, Pv PR2	Small acidic proteins, related to pea, potato, and parsley PR proteins and birch pollen allergen. Induced in cell cultures by fungal elicitor.	75
	Solanum tuberosum	λZAP	pSTH-2, pSTH-21	Similar to pea and parsley PR proteins. Induced by arachidonic acid in tuber tissue.	42
Others					
'Disease resistance response' genes	*Pisum sativum*	pUC9	pI49, pI176, pI206, etc.	Nine cDNA clones representing messages induced by fungal infection. At least 2 classes by sequence analysis.	20, 55
'Infection related mRNAs'	*Hordeum vulgare*		pRP1	0.88 kb mRNA induced by fungal infection.	28

[a] See refernce 50 for details of homologies of CHS cDNAs from different species, and details of the origin of CHS clones from *Antirrhinum majus*, *Hordeum vulgare*, *Magnolia liliiflora*, *Petunia hybrida*, *Ranunculus acer*, and *Zea mays*.

[b] PCR = polymerase chain reaction.

Table 2. Genomic clones for plant defence response genes

Enzymes of phytoalexin biosynthesis

Gene encoding	Source	Vector	Comments	References
Phenylalanine ammonia lyase	*Oryza sativa*	λEMBL3	Light-induced gene, small gene family	46
	Petroselinum crispum	λEMBL4	Family of at least 4 genes.	38
	Phaseolus vulgaris	λgt WES, 1059	Family of 3 different genes; differential regulation by infection, light, and developmental cues.	9
4-Coumarate: CoA ligase	*Petroselinum crispum*	λEMBL4	Two genes, both induced by elicitor and UV light.	14
Chalcone synthase	*Arabidopsis thaliana*	λEMBL4	Single gene, induced by high intensity light.	18
	Petroselinum crispum*	λEMBL4	Single gene but two alleles, one containing an insertional element. Both induced by UV light.	24
	Petunia hybrida*	λEMBL3	Family of at least 8 genes; two major genes expressed during development, 2 others UV-induced in seedlings.	29, 30, 31
	Phaseolus vulgaris	λ1059	Family of 6–8 genes, some linked; differential regulation by infection, light and developmental cues.	57
Chalcone isomerase	*Petunia hybrida**	λEMBL3	Two genes; differential expression (1 UV-induced)	72
Stilbene synthase	*Arachis hypogaea*	λEMBL3	Probably more than 3 genes; sequence similarity to chalcone synthase. Differential induction.	60, 35
HMG CoA-reductase	*Arabidopsis thaliana**	λEMBL4	Probably a single gene.	36, 4
	Lycopersicon esculentum			10
	Solanun tuberosum		Probably more than 3 genes.	67

Table 2. *Continued*

Gene encoding	Source	Vector	Comments	References
Hydroxyproline-rich glycoproteins				
Hydroxyproline-rich glycoprotein	*Daucus carota*[*]	λCharon 4A	Wound induced. More than one gene.	5
Hydrolases and 'PR' proteins				
Chitinase	*Arabidopsis thaliana*		Genes for acidic and basic chitinases.	58
	Phaseolus vulgaris	λEMBL4	Approximately 4 genes, at least 2 induced by ethylene.	3
'Chitinase-like'	*Solanum tuberosum*	λEMBL3	Endochitinase gene of cv Havana 425.	62
	Solanum tuberosum	λEMBL4	Two tandemly linked, wound-induced genes homologous to chitin-binding proteins.	66
'Pathogenesis-related' proteins	*Nicotiana tabacum*	λEMBL3	PR-la gene.	53
	Nicotiana tabacum	λCharon 32	PR-la gene, plus two other (silent?) PR genes.	8
	Nicotiana tabacum	λCharon 32	Two PR-S (thaumatin-like) genes, E2 and E22.	71
	Petroselinum crispum	λEMBL4	At least 2 genes. Rapid induction of PR1-1 by fungal infection. Little homology to tobacco PR-1 genes.	65
	Solanum tuberosum	λEMBL3	Gene corresponding to cDNA STH-21.	41
Others				
Thionins	*Hordeum vulgare*	λCharon 35	Multigene family, 50–100 members per haploid genome. All on chromosome 6.	2

198

References

1. Blyden, E. R., Doerner, P. W., Lamb, C. J., and Dixon, R. A. (1991). *Plant Mol. Biol.*, **16**, 167.
2. Bohlmann, H., Clausen, S., Behnke, S., Giese, H., Hillier, C., Reimann-Phillip, *et al.* (1988). *EMBO J.*, **7**, 1559.
3. Broglie, K. E., Gaynor, J. J., and Broglie, R. M. (1986). *Proc. Natl. Acad. Sci. USA*, **83**, 6820.
4. Caelles, C., Ferrer, A., Balcells, L., Hegardt, F. G., and Boronat, A. (1989). *Plant Mol. Biol.*, **13**, 627.
5. Chen, J. and Varner, J. E. (1985). *EMBO J.*, **4**, 2145.
6. Corbin, D. R., Sauer, N., and Lamb, C. J. (1987). *Mol Cell. Biol.*, **7**, 6337.
7. Cornelissen, B. J. C., Hooft van Huijsduijen, R. A. M., Van Loon, L. C., and Bol, J. F. (1986). *EMBO J.*, **5**, 37.
8. Cornelissen, B. J. C., Horowitz, J., van Kan, J. A. L., Goldberg, R. B., and Bol. J. F. (1987). *Nucleic Acids Res.*, **15**, 6799.
9. Cramer, C. L., Edwards, K., Dron, M., Liang, X., Dildine, S. L., Bolwell, G. P., *et al.* (1989). *Plant Mol. Biol.*, **12**, 367.
10. Cramer, C. L., Park, H. S., Denbow, C. J., Yang, Z., and Lacy, G. H. (1989). *J. Cell. Biochem. Suppl.*, **13D**, 316.
11. Cutt, J. R., Dixon, D. C., Carr, J. P., and Klessig, D. F. (1988). *Nucleic Acids Res.*, **16**, 9861.
12. Dalkin, K. and Dixon, R. A. (1990). *J. Cell Biochem. Suppl.* **14E**, 315.
13. De Loose, M., Alliotte, T., Gheysen, G., Genetello, C., Gielen, J., Soetaert, P., *et al.* (1988). *Gene*, **70**, 13.
14. Douglas, C., Hoffmann, H., Schulz, W., and Hahlbrock, K. (1987). *EMBO J.*, **6**, 1189.
15. Edington, B. V., Lamb, C. J., and Dixon, R. A. (1991). *Plant Mol. Biol.*, **16**, 81.
16. Edwards, K., Cramer, C. L., Bolwell, G. P., Dixon, R. A., Schuch, W., and Lamb, C. J. (1985). *Proc. Natl. Acad. Sci. USA*, **82**, 6731.
17. Epping, B., Kittel, M., Ruhnau, B. and Hemblen, V. (1990). *Plant Mol. Biol.*, **14**, 1061.
18. Feinbaum, R. L. and Ausubel, F. M. (1988). *Mol. Cell. Biol.*, **8**, 1985.
19. Fincher, B. G., Lock, P. A., Morgan, M. M., Lingelbach, K., Wettenhall, R. E. H., Mercer, J. F. B., *et al.* (1986). *Proc. Natl. Acad. Sci. USA*, **83**, 2081.
20. Fristensky, B., Horovitz, D., and Hadwiger, L. A. (1988). *Plant Mol. Biol.*, **11**, 713.
21. Fritzemeier, K-H., Cretin, C., Kombrink, E., Rohwer, F., Taylor, J., Scheel, D., *et al.* (1987). *Plant Physiol.*, **85**, 34.
22. Harker, C. L., Ellis, T. H. N., and Coen, E. (1990). *Plant Cell*, **2**, 185.
23. Hedrick, S. A., Bell, J. N., Boller, T., and Lamb, C. J. (1988). *Plant Physiol.*, **86**, 182.
24. Herrmann, A., Schulz, W., and Hahlbrock, K. (1988). *Mol. Gen. Genet.*, **212**, 93.
25. Hooft van Huijsduijnen, R. A. M., Cornelissen, B. J. C., van Loon, L. C., van Boom, J. H., and Bol, J. F. (1985). *EMBO J.*, **4**, 2167.
26. Hooft van Huijsduijnen, R. A. M., Kauffmann, S., Brederode, F. Th., Cornelissen, B. J. C., Legrand, M., Fritig, B., *et al.* (1987). *Plant Mol. Biol.*, **9**, 411.
27. Hooft van Huijsduijnen, R. A. M., van Loon, L. C., and Bol, J. F. (1986). *EMBO J.*, **5**, 2057.
28. Jutidamrongphan, W., Mackinnon, G., Manners, J. M., and Scott, K. J. (1989). *Nucleic Acids Res.*, **17**, 9478.

29. Koes, R. E., Spelt, C. E., and Mol, J. N. M. (1989). *Plant Mol. Biol.*, **12**, 213.
30. Koes, R. E., Spelt, C. E., Mol, J. N. M., and Gerats, A. G. M. (1987). *Plant Mol. Biol.*, **10**, 159.
31. Koes, R. E., Spelt, C. E., Reif, H. J., van den Elzen, P. J. M., Veltkamp, E. and Mol, J. N. M. (1986). *Nucl. Acids Res.*, **14**, 5229.
32. Kuhn, D. N., Chappell, J., Boudet, A., and Hahlbrock, K. (1984). *Proc. Natl. Acad. Sci. USA*, **81**, 1102.
33. Lagrimini, L. M., Burkhart, W., Moyer, M., and Rothstein, S. (1987). *Proc. Natl. Acad. Sci. USA*, **84**, 7542.
34. Laflamme, D. and Roxby, R. (1989). *J. Cell. Biochem. Suppl.*, **13D**, 325.
35. Lanz, T., Schroder, G., and Schroder, J. (1990). *Planta*, **181**, 169.
36. Learned, R. M. and Fink, G. R. (1989). *Proc. Natl. Acad. Sci. USA*, **86**, 2779.
37. Lois, A. F. and West, C. A. (1990). *Arch. Biochem. Biophys.*, **276**, 270.
38. Lois, R., Dietrich, A., and Hahlbrock, K. (1989). *EMBO J.*, **8**, 1641.
39. Maher, E. A. and Dixon, R. A. (1990). *J. Cell Biochem. Suppl.*, **14E**, 322.
40. Matsuoka, M., Yamamoto, N., Kano-Murakami, Y., Tanaka, Y., Ozeki, Y., Hirano, H., *et al.* (1987). *Plant Physiol.*, **85**, 942.
41. Matton, D. P., Bell, B., and Brisson, N. (1990). *Plant Mol. Biol.*, **14**, 863.
42. Matton, D. P. and Brisson, N. (1989). *Molec. Plant–Microbe Interactions*, **2**, 325.
43. Mehdy, M. and Lamb, C. J. (1987). *EMBO J.*, **6**, 1527.
44. Memelink, J., Hoge, J. H. C., and Schilperoort, R. A. (1987). *EMBO J.*, **6**, 3579.
45. Metraux, J. P., Burkhart, W., Moyer, M., Dincher, S., Middlesteadt, W., Williams, S., *et al.* (1989). *Proc. Natl. Acad. Sci. USA*, **86**, 896.
46. Minami, E., Ozeki, Y., Matsuoka, M., Koizuka, N., and Tanaka, Y. (1989). *Eur. J. Biochem.*, **185**, 19.
47. Mohnen, D., Shinshi, H., Felix, G., and Meins, Jr., F. (1985). *EMBO J.*, **4**, 1631.
48. Narita, J. O. and Gruissem, W. (1989). *Plant Cell*, **1**, 181.
49. Neale, A. D., Wahleithner, J. A., Lund, M., Bonnett, H. T., Kelly, A., Meeks-Wagner, D. R., *et al.* (1990). *Plant Cell*, **2**, 673.
50. Niesbach-Klosgen, U., Barzen, E., Bernhardt, J., Rohde, W., Schwarz-Sommer, Z., Reif, H-J., *et al.* (1987). *J. Mol. Evol.*, **26**, 213.
51. Payne, G., Ahl, P., Moyer, M., Harper, A., Beck, J., Meins, Jr., *et al.* (1990). *Proc. Natl. Acad. Sci. USA*, **87**, 98.
52. Pfitzner, U. M. and Goodman, H. M. (1987). *Nucleic Acids Res.*, **15**, 4449.
53. Pfitzner, U. M., Pfitzner, A. J. P., and Goodman, H. M. (1988). *Mol. Gen. Genet.*, **211**, 290.
54. Reimold, U., Kroger, M., Kreuzaler, F., and Hahlbrock, K. (1983). *EMBO J.*, **2**, 1801.
55. Riggleman, R. C., Fristensky, B., and Hadwiger, L. A. (1985). *Plant Mol. Biol.*, **4**, 81.
56. Ryder, T. B., Cramer, C. L., Bell, J. N., Robbins, M. P., Dixon, R. A., and Lamb, C. J. (1984). *Proc. Natl. Acad. Sci. USA*, **81**, 5724.
57. Ryder, T. B., Hedrick, S. A., Bell, J. N., Liang, X., Clouse, S. D. and Lamb, C. J. (1987). *Mol. Gen. Genet.*, **210**, 219.
58. Samac, D. A., Hironaka, C. M., Yallaly, P. E. and Shah, D. M. (1990). *Plant Physiol.*, **93**, 907.
59. Sauer, N., Corbin, D. R., Keller, B., and Lamb, C. J. (1989). *Plant Cell and Environ.*, **13**, 257.
60. Schroder, G., Brown, J. W. S., and Schroder, J. (1988). *Eur. J. Biochem.*, **172**, 161.
61. Shinshi, H., Mohnen, D., and Meins, Jr., F. (1987). *Proc. Natl. Acad. Sci. USA*, **84**, 89.

62. Shinshi, H., Neuhaus, J-M., Ryals, J., and Meins, Jr., F. (1990). *Plant Mol. Biol.*, **14**, 357.

63. Shinshi, H., Wenzler, H., Neujaus, J-M., Felix, G., Hofsteenge, J., and Meins, Jr., F. (1988). *Proc. Natl. Acad. Sci. USA*, **85**, 5541.

64. Somssich, I. E., Bollmann, J., Hahlbrock, K., Kombrinck, E., and Schulz, W. (1989). *Plant Mol. Biol.*, **12**, 227.

65. Somssich, I. E., Schmelzer, E., Kawalleck, P., and Hahlbrock, K. (1988). *Mol. Gen. Genet.*, **213**, 93.

66. Stanford, A., Bevan, M., and Northcote, D. (1989). *Mol. Gen. Genet.*, **215**, 200.

67. Stermer, B., Edwards, L. A., Edington, B., and Dixon, R. A., (1991). *Physiol. Mol. Plant Pathol.* (in press).

68. Swegle, M., Huang, J-K., Lee, G., and Muthukrishnan, S. (1989). *Plant Mol. Biol.*, **12**, 403.

69. Takeuchi, Y., Yoshikawa, M., Takeba, G., Tanaka, K., Shibata, D., and Horino, O. (1990). *Plant Physiol.*, **93**, 673.

70. Tanaka, Y., Matsuoka, M., Yamanoto, N., Ohaski, Y., Kano-Murakami, Y., and Ozeki, Y. (1989). *Plant Physiol.*, **90**, 1403.

71. van Kan, J. A. L., van de Rhee, M. D., Zuidema, D., Cornelissen, B. J. C., and Bol, J. F. (1989). *Plant Mol. Biol.*, **12**, 153.

72. van Tunen, A. J., Hartman, S. A., Mur, L. A., and Mol, J. N. M. (1989). *Plant Mol. Biol.*, **12**, 539.

73. van Tunen, A. J., Koes, R. E., Spelt, C. E., van der Krol, A. R., Stuitje, A. R., and Mol, J. N. M. (1988). *EMBO J.*, **7**, 1257.

74. Walter, M. H., Grima-Pettenati, J., Grand, C., Boudet, A. M., and Lamb, C. J. (1988). *Proc. Natl. Acad. Sci. USA*, **85**, 5546.

75. Walter, M. M., Lin, J-W., Grand, C., Lamb, C. J., and Hess, D. (1990). *Mol. Gen. Genet.*, **222**, 353.

A2

List of suppliers

Agar Aids Ltd, PO Box 101, Hemel Hempstead, Herts, UK.

Aldrich Chemical Co., 940 West Saint Paul Avenue, Milwaukee, WI 53223, USA; The Old Brickyard, New Road, Gillingham, Dorset, SP8 4JL, UK.

Alltech, Deerfield, ILL, USA.

Amersham International, Amersham Place, Little Chalfont, Buckinghamshire HP7 9NA, UK; 2636 S. Clearbrook Drive, Arlington Heights, IL 60005, USA.

Amicon Corporation, 17 Cherry Hill Drive, Danvers, MA 01923, USA; Upper Mill, Stonehouse, Gloucester GL10 2J, UK.

Applied Biosystems Inc., 850 Lincoln Center Drive, Foster City, CA 94404, USA; Kelvin Close, Birchwood Science Park North, Warrington, Cheshire WA3 7PB, UK.

Bayer, UK Supplier, Dalgetty Agriculture Ltd, Green Lane West, Rackheath, Norwich, NR13 6N4, UK.

BDH, Broom Road, Poole, Dorset, BDH12 4NN, UK.

Beckman Instruments Ltd, Progress Road, Sands Industrial Estate, High Wycombe, Buckinghamshire, HP12 4YL, UK.

Bio 101 Inc., Box 2284, La Jolla, CA 92038–2284, USA; Stratech Scientific Ltd., 61/63 Dudley Street, Luton, Bedfordshire, LU2 0NP, UK.

Bio-Rad, 1414 Harbour Way South, Richmond, CA 94804, USA; Claxton Way, Watford Business Park, Watford, Hertfordshire, WD1 8RP, UK.

Biosupplies, PO Box 835, Parkville, Australia.

Boëhringer-Mannheim GmbH, Postfach 310120, D-6800 Mannheim 31, Germany; PO Box 50816, Indianapolis, IN 46250, USA; Bell Lane, Lewes, Sussex, BN7 1LG, UK.

BRL. See Gibco-BRL.

British Biotechnology Ltd, 4–10 The Quadrant, Barton Lane, Abingdon, Oxon, OX14 3YS, UK.

Brownlee Laboratories, Santa Clara, CA 95050, USA.

BTX, 3742 Jewell Street, San Diego, CA 92109, USA.

Calbiochem, 10933 North Torrey Pines Road, La Jolla, CA 92037, USA; Novobiochem (UK) Ltd, Freepost, Nottingham NG7 1BR, UK.

Cambio, 34 Millington Road, Newnham, Cambridge, CB3 9HP, UK.

Cambridge Bioscience Ltd, 25 Signet Court, Stourbridge Business Centre, Cambridge, CB5 8LA, UK.

Cen Saday, Molecules Marquees, Gif-Sur-Yvette, France.

Cetus. See **Perkin Elmer Cetus.**
Chempack Products Ltd, Geddings Road, Hoddesdon, Hertfordshire, UK.
Ciba Corning Diagnostic Corporation, Oberlin, Ohio, USA.
Costar (see NBL).
Difco Laboratories Ltd, PO Box 14B, Central Avenue, East Moseley, Surrey, England.
Dionex Ltd, Camberley, Surrey, GU15 2PL, UK.
DuPont Company, Biotechnology Systems Division, BRML, G-50986, Wilmington, DE 19898, USA; Wedgewood Way, Stevenage, Hertfordshire, SG1 4QN, UK.
Fisher Scientific, Pittsburgh, PA, USA; Zurich, Switzerland.
FMC Bioproducts, 5 Maple Street, Rockland, ME 04841-2994, USA; Flowgen Instruments Ltd, Broad Oak Enterprise Village, Sittingbourne, Kent, ME9 8AQ, UK.
ICN/FLOW, High Wycombe, Bucks, HR13 7DL, UK.
Genetic Research Instruments Ltd, Gene House, Dunmow Road, Felsted, Dunmow, CM6 3LD, UK.
Gibco-BRL, Grand Island, NY, USA; PO Box 35, Trident House, Renfrew Road, Paisley, PA3 4EF, UK.
IBF Biotechnics, Columbia, MD, USA.
Invitrogen, San Diego, CA 92121, USA (see also British Biotechnology Ltd).
Jenssen Life Sciences and J. W. Scientific Products Merck Darnstadt, Germany.
J.R. Scientific (see Seralay UK Ltd).
Karlan Chemical Corporation, 23875 Madison Street, Torrance, CA 90505, USA.
Kinematica UK, Philip Harris Scientific, 618 Western Avenue, Park Royal, London W3 0TE, UK.
Kodak (Eastman Kodak), Acorn Field Road, Knowsley Industrial Park North, Liverpool, LS3 72X, UK; Rochester, New York 14650, USA.
LEP Scientific, Sunrise Parkway, Linford Wood East, Milton Keynes, Bucks, MK14 6QF, UK.
Marine Colloids, FMC Corporation, Bioproducts Division, 5 Maple Street, Rockland, ME 04841, USA; 1 Risingevej, DK-2665 Vallensbaek Strand, Denmark.
Millipore Corp., 80 Ashby Road, Beford, MA 01730, USA; The Boulevard, Blackmoor Lane, Watford, Hertfordshire WD1 8YW, UK.
MJ Research, Kendall Square, Box 363, Cambridge, MA 02142, USA.
New England Biolabs, 32 Tozer Road, Beverley, MA 01915, USA; Postfach 2750, 6231 Schwalbach/Taunus, FRG; CP Laboratories, PO Box 22, Bishop's Stortford, Herts, CM23 3DM, UK.
New England Nuclear. See **DuPont.**
Northrop-King Co, PO Box 959, MIN, USA.
Northumbrian Biological Ltd, South Nelson Industrial Estate, Cramlington, Northumberland, NE23 9HL, UK.
Novo Biolabs, Novo Industri A/S, Novo Allé, Ak-2880 Bagsvaerd, Denmark.

Onozuta, Karlan, Torrance, CA, USA (see also Yakult Honsha Ltd).

Operon Technologies, Almeda, CA, USA.

Oxoid Ltd, Wade Road, Basingstoke, RG24 0PW, Hants, England.

Parr Instrument Co., Molline, Illinois, USA.

Perkin Elmer Cetus, Main Avenue, Norwalk, CT 06859-0012, USA; Postfach 101164, 7770 Ueberlingen, FRG; Maxwell Road, Beaconsfield, Buckinghamshire HP9 1QA, UK.

Polysciences, Paul Valley Industrial Park, Warrington, PA 18976, USA.

Pharmacia LKB Biotechnology, AB, S-75182 Uppsala, Sweden; 800 Centennial Avenue, Piscataway, NJ 08854, USA; Southampton SO1 7NS, UK.

Promega Corporation, 2800 South Fish Hatchery Road, Madison, WI 537-5305, USA; Episcon House, Enterprise Road, Chilworth Research Centre, Southampton, SO1 7NS, UK.

Raymond A. Lamb, 6 Sunbeam Road, London, NW10 6YL, UK.

Sartorius Ltd, Belmont, Surrey SM2 6JD, UK.

Seishin Pharmaceuticals Co Ltd, 4.13, Koamicho, Nihoubashi, Tokyo, Japan.

Sera Lab UK Ltd, Crawley Down, Sussex, RH10 4FF, UK.

Serva Feinbiochemica GmbH & Co., D-6900, Heidelberg 1, PO Box 105260.

Stewart Plastics, Purley Way, Croydon, UK.

Sigma Chemical Co., PO Box 14508, St. Louis, MS 63178, USA; Fancy Road, Poole, Dorset BH17 7NH, UK.

Spectrum Medical 44 Upper Northgate Street, Chester CH1 4EF, UK.

Stratagene Ltd, Cambridge Innovation Centre, Cambridge Science Park, Milton Road, Cambridge CB4 4GF, UK; 11099 North Torrey Pines Road, La John, CA 92037, USA.

Supec Inc., Bellefonte, PA, USA.

Tetko Inc., New York, USA.

US Biochemicals (see British Biotechnology Ltd).

Whatman Biosystems Ltd, Springfield Mill, Maidstone, Kent ME14 2LE, UK; 9 Bridewell Place, Clifton, NJ 07014, USA.

Weck & Co. Inc., Research Triangle Park, NC, USA.

Yakult Honsha Co. Ltd, Medicine Department, Enzyme Division, 1-1-19, Higashi-Shinbashi, Minatokv, Tokyo 105, Japan.

Zeiss, Welwyn Garden City, Hertfordshire, AL7 1LU, UK.

Index

Index

Index

Contents of Volume II

SECTION 3.
DEFENCE RESPONSES (GENE PRODUCTS)

215

Contents of Volume II

SECTION 4. ELICITORS

SECTION 5. SIGNAL TRANSDUCTION PATHWAYS

SECTION 6. DISEASE RESISTANCE GENES